BOTANICAL TREASURES

Multi-Use Plants for Renewable Resources
and a
Nature-Based Economy

by

Joshua Smith

Mud City Press
Eugene, Oregon

Botanical Treasures

Published by
Mud City Press
http://www.mudcitypress.com
Eugene, Oregon

The photo on the cover of this book was taken by Thomas Checkley. The photo and all of the plant illustrations within are sourced at the back of the book .

ISBN-978-0-9830045-8-5

Printed in the United States

To Mother Earth

Table of Contents

Author's Note...viii

Acknowledgments..ix

Introduction..xi

Achillea Lanulosa (Yarrow)...1

Aleurites Species (Candle Nut, Tung Oil Tree)..................15

Allium Sativum (Garlic)...22

Asclepias Species (Milkweed)..40

Cucurbita Foetidissima (Buffalo Gourd)...........................61

Eucommia Ulmoides (Hardy Rubber Tree).......................68

Fraxinus Species(Ash)..77

Gledistsia Triacanthos (Honey Locust)............................89

Lespedeza Bicolor (Bush Clover)....................................100

Myrtus Communis (Myrtle)..104

Origanum Vulgare (Oregano)...110

Phragmites Communis (Common Reed)...............................122

Prosopis Species (Mesquite)..132

Rosa Species (Rose)..150

Sambucus Species (Elderberry)......................................176

Trachycarpus Fortunei (Hemp Palm)..................................201

Typha Species (Cattail)..205

Yucca Species (Yucca)..225

Ziziphus Jujuba (Jujube)...244

Appendix A. Hardiness Map for the U.S. and Canada.........258

Appendix B. Ecological Plant Culture...............................259

Glossary...270

Bibliography...277

Illustration Sources..285

The Author..288

Author's Note

Many of the plants profiled in this book include a segment on their medicinal use. The intention of these segments is to report on the work of ethnobotanists, medical scientists, researchers, and other healthcare professionals who have studied these plants. Not being a healthcare professional myself, I am not recommending self-diagnosis, prescribing medicine, or treating the ills discussed. Indeed, I strongly argue against it. The purpose of the medical segments is informational. Hopefully it will be useful to healthcare professionals specializing in nature-based medicine and nutrition.

Acknowledgments

First I must express my deepest gratitude to my friend, publisher, and editor Dan Armstrong. His guidance has made this a far better book than I could have hoped for. Thanks also to Tyler Armstrong, his contribution is greater than he knows. Then there is the man who has been a great inspiration to me, a priceless teacher and a real friend. That would be Bill Mollison, whose name is synonymous with permaculture design, one of the most significant developments of our time.

I would also like to thank Thomas Checkley for the photo that graces the cover of this book. And a special thanks goes to Gaea Yudron, James A. Duke, Greg Carey, Brenda Amick, and Harry MacCormick for reading the manuscript and offering their comments.

Introduction

There is a rare, exquisite gem in our solar system, an egg-shaped rock with a fire in her belly. She's the crown jewel among Sol's offspring. This rock gave birth to all life within our biosphere and has provided for us all ever since—which is why from the earliest times we have called her Mother Earth. It would seem wise to treat our divinely mysterious mother as sacrosanct.

Wisdom, however, is not necessarily on the agenda of our world leaders. In this era of critical global issues, such as the unprecedented rate of extinction of many plant and animal species, the pollution and rapid loss of our fresh water supplies, or the threat of nuclear war, none stand out as glaringly as climate change—because if we don't vastly diminish the burning of fossil fuels, transform our economy, and very deliberately change the way we live, then we, the human race, and nearly all life on Earth, will suffer the consequences.

The period in which we are currently living is more perilous than any other in history, and we only have ourselves to blame. What is demanded of us now is a radical transformation; what might be called an evolutionary leap in consciousness. Every global extinction that has occurred on planet Earth has had its survivors. These were species with the ability to adapt to catastrophic change. Today's evolving climate and the ongoing sixth great extinction are not going to allow us time to adapt physically; however, we are capable of conscious evolutionary adaptation. One facet of this is learning to live in accord with the entire community of life—a task that means nothing less than reinventing society so that it is co-evolutionary with the rest of nature.

Management of our natural resources in both an efficient and sustainable manner is vital to such a vision. Part of that is understanding what nature's botanical treasures have to offer and how to cultivate them. *Botanical Treasures* provides this information for plants from nineteen genera while also hoping to bring attention to a form of nature-based renewable economics that has been all but overlooked in this age of technology.

The plants that are profiled in this book have the potential to produce a wide range of raw materials for a variety of products and uses. What's more, some of these plants can adapt productively to disturbed or marginal lands, preserving valuable farmland for less tolerant crops. Some contribute valuable functions to agroecology and ecological rehabilitation. Some can supply us with substitutes for fossil fuels or other petrochemical products like rubber, plastic, adhesives, caulk, insulation, cellophane, surfactants, varnish, wax, waterproofing, shampoo, and cosmetics. Some can provide fiber for anything from textiles to baskets to brooms. Some can purify wastewater, be used to treat sewage, and in some cases, remove heavy metals from mine tailings. Many of them are medicinal, and in some cases outstandingly so. And of course, many of these botanical treasures are edible, often quite tasty, and in some cases exceptionally nutritious.

The potential benefits of a plant-based, environmentally propitious economy are considerable. If natural resources are renewable and are also utilized in an ecologically sustainable manner, they become important tools for confronting the manifold challenges of global warming. (See Appendix B. Ecological Plant Culture, page 259.)

Perhaps the time for the eco-entrepreneur has come. Certainly an acceptable income can be derived from the resources offered by these botanical treasures. In this era of environmental concern and worker disenfranchisement, wouldn't it be vastly empowering to produce a variety of products in ways that not only heal the planet but also create jobs? Jobs one can take pride in because they contribute to something truly meaningful! Hopefully the material in this book will help point the way to this necessary transformation.

Joshua Smith, July 18, 2016

ACHILLEA LANULOSA
[Syn. A. millefolium, A.M. Var Lanulosa]
(Wild Yarrow, Common Yarrow, Milfoil Plumajillo)
[Asteraceae Family]

Wild yarrow is native to the old world and the new. A few relatively small differences between the species have been identified, however, they readily cross pollinate and are hard to tell apart. Some botanists now treat them as one species. Regardless of name, their medicinal and other uses are considered interchangeable.

Wild yarrow is a perennial herb that tends to form a spreading mat. The aromatic foliage is finely cut and ferny, each leaf being 2 to 8 inches long. The flowers are borne in flattop or dome shaped clusters, 2 to 4 inches across, on stems 3 inches to 3 feet tall. The flowers are typically white to off white, but occasionally pink to red, and bloom from May to September.

Culture:

Yarrow prefers full sun to part shade, and tolerates almost any soil type as long as it drains fairly well. It is relatively drought tolerant, depending on ecotype. Yarrow is at its best, however, in moist, but not wet soil, and is most often found on well-drained sandy to gravely loams with dry to moderately moist soil.

Yarrow is easy to grow from seed and propagates well from division. Space transplants 8 inches to 1 foot apart, and

1

Achillea Millefolium L.

they will fill in fast. To rejuvenate an old patch take some divisions out of it. It tolerates overgrazing by livestock or wildlife.

Yarrow seeds are highly viable and should not even be covered to allow for good germination. Mulch them very lightly, with a bit of straw, just enough to minimize soil evaporation. One caution, keep them well away from young trees. When the trees are older, yarrow will be a valuable companion, but while young, they can compete with these baby trees fairly aggressively.

Ecotypes differ in their tolerances, particularly their adaptation to cold. Plants from the mountains are a poor choice for planting at low altitudes, and those from the lowest altitudes will increasingly struggle as they are planted at ever-higher altitudes. For the sake of surviving an overheating planet, however, it is best to select yarrow ecotypes from warmer zones and plant them further north or higher in altitude.

Varieties:

Achillea Millefolium Var. Proa was bred for a high essential oil content. The essential oil itself is a rich source of the anti-inflammatory agent proazulene. This variety's increased medicinal potency, productivity, and ease of culture recommend it for commercial cropping.

Food:

Few people today would think of wild yarrow as a food despite its long history as such. Nevertheless wild yarrow is not just edible, it is quite nutritious. Containing more than 120 compounds, wild yarrow as a food also serves as a preventative medicine.

Leaves

Yarrow's tender young aromatic leaves, when finely chopped, can be included in fresh salads, though they tend to stick to the top of the mouth and in the throat. They can be steamed or boiled, seasoned, and served as a vegetable dish, or better yet, you can combine them with other greens.

When the leaves are dried and crushed, they can serve as a seasoning or a base for soup. The herb is a nice addition to soups, stews, and sauces.

The foliage has also been used as a substitute for hops in brewing beer. It has just the right amount of bitterness and produces a strong beer with a distinctive pleasing flavor.

Flowers and Seeds

When the leaves, flowers, or seeds are dried they make a delightful tea when a dash of honey is added. The strongly aromatic flowers can be dried and ground to serve as a flavoring, but go easy. The flowers' essential oil is used commercially as a flavoring, primarily in soda pop, or as a hops substitute for Swedish beer or other alcoholic beverages. It is best to avoid consuming yarrow daily, except when used medicinally, because of its high tannin content and the potential to damage mucus membranes.

Medicine:

Functions

In his book *Medicinal wild Plants of the Prairie*, Kelly Kindscher informs us that at least fifty-eight First Nation tribes used yarrow to treat a variety of ills. It was also used in Europe and China, often in much the way it was used by our First Nation peoples.

The vitamins and minerals that make up yarrow's nutritional profile are many. It is extra rich in chromium, rich in vitamin C, vitamin B1 (thiamine), vitamin B2 (riboflavin), choline (a member of the vitamin B complex), potassium, and selenium, but also contains calcium, manganese, magnesium, phosphorous, silicon, sulfur plus protein, and dietary fiber.

Yarrow also possesses a rich complex of phytochemicals which are responsible for its use as a medicinal. Make no mistake, this is a very important herb.

Yarrow's essential oil contains chamazulene, sabine, I-8 cineol, A and B pinene, camphor, bornyl acetate, borneol, and several other aromatics, including the sequiterpene lactones achillicin, achillin, achillifolin, millifin, and millifolid. It also contains several flavonoids (apigenin, luteolin, isorhamnetin, rutin), phenolic acids (caffeic, salicylic, and alkaloids), betonicine, stachydrine, achiceine, moschatine, and trigonelline.

Yarrow acts as an analgesic, an antibacterial, an antibiotic, and an anticatarrhal, containing over a dozen anti-inflammatory compounds as well as numerous antispasmodic, antithrombotic, carminative, diaphorietic, diuretic, expectorant, and febrifuge compounds, not to mention several hemostatic, hepatic, hypoglycemic, hypotensive, stimulant, and vulnerary compounds.

Circulatory System

As a blood tonic yarrow tea promotes good circulation and tones the blood vessels. Its high chromium content contributes to the spleen's ability to eliminate worn out red blood cells. Chromium is also important for maintaining blood sugar metabolism and lowering blood pressure. Clinical studies have shown that chromium helps reduce elevated levels of blood glucose and improves glucose tolerance.

Chromium is reputed to be an anti-aging agent that reduces LDL cholesterol, raises HDL cholesterol, and lowers triglyceride levels. Chromium is of particular value to diabetics. Yarrow's chromium content, working together with its alkaloids, may lower the need for insulin injections by diabetes mellitus sufferers. Other functions attributed to chromium include weight loss for the overweight and body mass increase for the lean. It treats both glaucoma and acne as well.

As a hemostatic and vulnerary herb, yarrow tea can slow or stop bleeding, and in some cases even control hemorrhages. Its vulnerary action then goes to work healing the wound. Although the fresh herb works best, both the dried foliage and flowers will do the job. In most cases, though not all, tea is taken for internal bleeding, and a poultice made from the leaves is used for external bleeding.

It can be taken internally to treat hemorrhages of the lungs or kidneys, and can help heal bleeding stomach ulcers. For hemorrhages of the intestines, yarrow tea is given as an enema. It is also mildly helpful for both bleeding and non-bleeding hemorrhoids. For nosebleeds take a fresh leaf, roll it up, and stick it in the problem nostril.

Women's Health

Yarrow tea helps regulate menstruation. It can reduce excessive menstrual flow and stimulate flow when obstructed. The tea is also taken for postpartum bleeding. In fact, yarrow is an herb that is considered beneficial to the entire female reproductive system.

Yarrow is recommended by Germany's Commission E for the relief of menstrual cramps. The numerous antispasmodic compounds contribute to relief from the pain.

In premenstrual syndrome (PMS), where pain is the primary symptom, yarrow tea and/or an herbal bath infused with yarrow are used by many European women.

If used judiciously as a tea, it can help stabilize the blood supply to the mucous membrane that lines the uterus. It has treated uterine hemorrhages, and it can reduce hot flashes during menopause.

Yarrow tea has been recommended to females of all ages except during pregnancy. That said, the women of the Blackfoot Nation took the tea once their water broke. They believed yarrow sped up delivery, helped expel the afterbirth, and promoted the production of breast milk.

Yarrow helps tone the uterus. A poultice of the leaves can be applied to sooth oversensitive breasts.

Yarrow is also used to help eliminate leucorrhea, a condition in which a white odorless mucopurulent vaginal discharge occurs, often caused by infection during ovulation, pregnancy, or excessive sexual activity.

Gastro-intestinal System

Yarrow is a major anti-inflammatory herb due to its wealth of sequiterpene lactones and its potent bioflavonoids, with likely support from yarrow's alkamides content. Yarrow tea has been used in the treatment of numerous inflammatory conditions, particularly those of the intestinal tract.

Yarrow has been a component for treatment of ulcerative colitis (UC), a chronic disease of the colon, and Crohn's disease, another serious inflammatory bowel disease. Indigestion, gastritis, heartburn, stomachaches, muscle pain, sprains, and intestinal flu all benefit from yarrow's carminative and antispasmodic actions as well as its anti-inflammatory compounds.

Although yarrow acts as a mild laxative, a water concentrate of the whole plant is used to control dysentery. Yarrow can also stimulate the appetite, and for that reason, it has been used as a component of anorexia nervosa treatment.

Yarrow's flower tea or tincture stimulates the digestive system to produce digestive enzymes and promotes the flow of bile from the gall bladder. The tea or tincture can promote the removal of excess acids from the intestines and undigested fats from the gallbladder. The tea can also eliminate excess buildup of gas. A fresh infusion makes an effective stomach tonic.

Cystitis is an acute or chronic inflammatory infection of the urinary bladder. Care in diagnosis is important because its symptoms are similar to sexually transmitted diseases, vaginitis, and an irritated urethra. Yarrow's anti-inflammatory, anti-microbial, and diuretic action can be a highly effective treatment for cystitis particularly given fresh.

Mucus Membranes

Yarrow's anticatarrhal action can soothe and stimulate the mucus membranes and provides pain relief from acute inflammation of mucus membranes of the nose or throat. As a decongestant it also effectively eliminates excessive mucus. Yarrow's alkamide content contributes much to its localized anesthesia for the nose and throat as well as providing some toothache relief. Alkamides are also considered responsible for stimulating the mucus membranes.

Colds, coughs, flu, fever, and chills are often successfully treated with yarrow. As an antispasmodic yarrow gives relief and may even prevent muscle cramps and spasms, particularly those that occur in the smooth muscles of the intestines or uterus. Yarrow's antispasmodic action is a result of its flavonoids and coumarins.

Liver Functions

Yarrow is also a valuable hepatic herb. It regulates liver functions and is an important aid to liver health and vitality.

Almost all liver conditions are improved by yarrow, including the stimulation of the flow of bile. Since the liver stores vitamin A, it is probable the tonic would liberate some of the vitamin A to supply other parts of the body with this antioxidant vitamin.

Joint Pain
Yarrow flowers' essential oil is an effective topical treatment for inflamed joints (antirheumatic). The glycol-proteins in yarrow's protein-carbohydrate complex appear to repair damaged tissue and swollen joints by forming aggregates around the inflamed or wounded area. Add the essential oil to the bathwater and receive a good pain relieving soak, or use the diluted essential oil or flowers as a poultice.

Burns
Wild yarrow was highly regarded by some First Nation healers as a treatment for burns. Typically a cold infusion of the whole plant was used. The Zuni pulverize the whole plant, add the results to cold water, and apply it directly to the burn. Zunis have secret societies whose ceremonies have included fire dancing and putting hot coals in their mouths. For protection they would chew on yarrow's roots and flowers. This mass could then be applied to the feet and legs as well.

Harvest and Preparation
It is believed that yarrow's medicinal effectiveness is greatest when it is harvested just as the flower clusters first appear. When harvesting the flower clusters, it is recommended to keep some stem and foliage with each cluster to maximize their potency and longevity.

Foliage is best dried on screens in a shaded, well-ventilated location. Flower clusters can be dried on tarps in partial shade (cover them at night) or dry them as you would the foliage. Store dried material in sealed glass jars in cool, dark, dry locations. Storage life is about one year.

For a tasty healing tea, infuse one or two teaspoons of the dried herb in a cup of boiling water and steep ten to fifteen minutes. The dried herb is more water-soluble than fresh and also tastes better. If taken as a tincture, use 2 to 4 milliliters or up to twenty drops three or four times daily.

Considerations and Warnings:

People allergic to ragweed or other members of the asteraceae family might have a dermatological reaction to yarrow's sesquiterene lactones. This, however, has no ill effect when taken internally. It is quite possible that extended use could result in mucus membrane damage. Sensitive individuals may experience photosensitization when yarrow is used daily for prolonged periods. Using the cautionary principle, yarrow should not be taken internally while pregnant. Yarrow's seeds possess a narcotic glycol-alkaloid. It has been reported that consuming the seeds can lead to a hallucinogenic effect.

Agricultural Uses:

Beneficial Insect Habitat

Beneficial insects are insects that help control insects that are agricultural pests. They do this either as parasites or as predators of the pests. On the farm or in the forest (really just about anywhere) yarrow's flowers provide a good protein source for beneficial insects. At least 1,000 or more of these tiny insects have been counted over the course of one hour visiting a single flower cluster. Beneficial parasitic wasps (they do not bite), Hoover flies, and ladybugs (lady beetles) are among those insects fond of yarrow. Some of the pests they have controlled include corn borers, saw flies, tomato hornworms, and numerous larva stage pests.

When it comes to pesticides, it is important to realize that beneficial insects are at least as vulnerable as the pests they would otherwise control. Our good bugs do not develop insecticide resistance nearly as well as the pests. Even some of the botanical insecticides can kill them, and should only be used as a last resort.

Biodynamics

In the 1920s, Rudolf Steiner, the founder of biodynamic agriculture, described yarrow as a marvel, indeed a miraculous creation. According to Steiner, it can regulate potassium processes in plants. Its flower clusters are standard ingredients in the biodynamic compost starter preparation #502. It is exceptionally adept at using its sulfur content to beneficially unify all its other constituents. In addition to triggering the biological processes of the compost pile, biodynamic preparation #502 is also recommended for the even ripening of tomatoes, as well as a cure for greenback.

Steiner said that as a companion plant wild yarrow imparts to its neighbors an increased ability to withstand adverse conditions.

It is a long held belief that when yarrow is grown in the company of aromatic herbs, their essential oil aromas are increased. However, when yarrow is grown alongside stinging nettles, nettles are the more aggressive of the two and on occasion must be dug out to protect the yarrow. In other cases, yarrow itself may be the aggressor.

Soil Fertility

Wild yarrow is a hyperaccumulator of minerals that are submerged deep in the soil. Wild yarrow will mine the subsoil for calcium, potassium, or silicon, but only if their presence is low in the surface soil layers. When the plant drops its foliage for winter dormancy or simply dies, those minerals become available to other plants through decomposition. Wild yarrow makes a good permanent ground cover in orchards once the trees are well established. It is best mixed with nitrogen fixers, other hyperaccumulators or accumulators, and plants that support pollinators or other beneficial insects.

Benefits for Vegetables
In their book *The Organic Method Primer—Special Edition*, the Rateavers say that wild yarrow improves the growth of vegetables in its vicinity due to an unknown compound.

Other Uses:
Erosion Control
Yarrow is excellent at controlling erosion. Its extensive root system will bind eroding soil on slopes, berms, cuts, and gullies. Its roots anchor it to the soil and it spreads aggressively.

Fire Barrier
Yarrow is fire resistant. With adequate amounts of moisture, it can survive just about anything short of molten lava. For this reason, yarrow makes a good ground cover for the protection of homes and forests. It is a good choice for firewise landscapes, as an ornamental ground cover, or a traffic-free lawn substitute.

Cockroach Control
The I-8 cineol found in yarrow's essential oil has demonstrated potential as a cockroach repellent.

Hair and Skin Products

While the Europeans found numerous applications for Achillea millefolium, the First Nation people of this continent had many uses for Achillea lanulosa before the arrival of the colonists. One specific use was skin and hair care.

Today yarrow is a common ingredient in nature-based commercial hair and skin products. It is particularly popular in shampoos because of its renown for preventing baldness. The flowers are preferred for hair care products because they give hair sheen. The tea also serves well as a hair rinse.

Yarrow is much valued for soothing, cooling, and firming the skin. Its delightful aromatic properties add to its value as an ingredient to skin lotion or other aromatics like incense, perfume, cologne, and bath powders.

Yarrow added to bath water is deeply soothing and leaves one refreshed. In his book *Medicinal Plants of the Pacific West*, Michael Moore said that the complexion of his scalp was much improved by simmering the flowers in water and placing his head—the portion that had stopped producing hair—in the steam.

Floral Arrangements

Yarrow's flower clusters are often used in floral arrangements. They are long lasting whether fresh or dry, and they offer a nice contrast to more vivid flowers. To keep the clusters open when drying, the cluster stems are cut to the desired length, stuck in sand, and covered with a paper bag. Better yet just hang them upside down in a cool, dry location.

Fish Preparation

When drying salmon, the people of southern Oregon's Klamath tribe stuffed each salmon with wild yarrow to accelerate the time it takes for them to thoroughly dry.

Ecological Functions:

Yarrow is as nutritious to its plant companions as it is to people and animals. It is known as a pioneer species, which is one that increases soil fertility in disturbed areas so other plants can follow. It improves the soil by capturing eroding sediment and organic matter and by mining the soil for minerals and making them available to other plants .

Wild yarrow is a very desirable nest building material in the bird world. Researchers at Wesleyan University appear to have discovered one reason birds favor yarrow. They seem to know that yarrow's antimicrobial action suppresses bacteria that degrade fledglings' feathers.

Native Range and Habitat:

A fair amount of controversy surrounds wild yarrow's presence in North America. Achillea millefolium was once thought to be a plant that was indigenous to Europe and Western Asia, and then it invaded the continental United States. However, the discovery of a native species, Achillea lanulosa, in the western United States suggests that it was well established in America long before the Europeans arrived. Some taxonomist realized that the perceived differences of the two plants were insufficient to classify them as two different species, so lanulosa was considered a subspecies of millefolium. Today they are often treated as the same species period.

So where does that leave us regarding wild yarrow's natural range? Wild yarrow is found in every state except Hawaii. It grows from sea level to subalpine zones, occasionally even above timberline. Wild yarrow occurs in such a wide range of habitats it seems ubiquitous. Achillea millefolium is found throughout Europe primarily on grasslands but also on wastelands and cultivated ground.

ALEURITES SPECIES
[Euphorbiaceae Family]

Aleurites are evergreen trees that produce seeds which contain commercial grade oil. The seeds are found in their 2 to 3 inch diameter fruits. Both species, *Aleurites moluccana* and *Aleurites fordii*, are considered hardy to zone 9 (See Appendix A. Hardiness Map for the United States and Canada, page 258).

Aleurites Moluccana
(Candlenut, Varnish Tree, Kukui, Otaheite Walnut, Jophal, Kemiri, Buah Keras, etc.)

Candlenut is a fast growing tree, 35 to 60 feet tall and about as wide. The flowers are small and white, borne in panicled cymes at the ends of the branches. The foliage is a light grayish green; at a distance the tree appears silver. Its olive green, 2 to 2½ inch fleshy fruit contains 1 or 2 large seeds (nuts). The bark is also gray-green.

Culture:
The candlenut prefers sun or light shade and likes acid soil with regular watering. It is resistant to light frost and does well along the Southern California coastline and the coastal thermal belt to the north. It is well-adapted to marginal conditions and has been naturalized in Hawaii and other tropical regions. It will grow in zone 9, but is at its best in zone 10. Propagation is possible by sowing seeds in place or by transplanting seedlings when they are 1 foot tall.

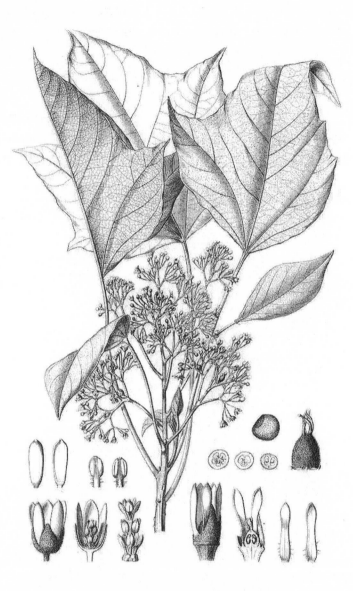

Aleurites Molucanna

Food:

Candlenut seeds can be baked or roasted, but they should be eaten in moderation due to their laxative effect. An edible cooking oil is liberated from the roasted seeds. The cooked seeds are often used to make condiments, such as flavoring for sauces or a thickener for curries.

The nuts are highly resistant to decay, even after a couple years on the ground. Rancid nuts are easily recognized because they float rather than sink when placed in water.

Medicine:

The oil is used in Indonesia to treat constipation and calluses. Its leaves are boiled and the nut is pulped for treatment of swollen joints, headaches, and ulcers. Candlenut's bark is used to alleviate dysentery.

Warnings and Considerations:

Never eat raw, uncooked candlenuts because they can be violently purgative.

Agricultural Uses:

Feed

The byproduct of candlenut oil production can be formed into an oil cake, fermented, and fed to livestock.

Fertilizer

The oil cake can also be used as a natural fertilizer.

Bee Forage

The trees supply habitat for honey bees.

Others Uses:

Drying Oil

Candlenut seeds contain a drying oil considered superior to linseed oil. Each tree can yield up to 200 pounds of seed, or 5 to 8 tons per acre. The oil is often used in paints, including artists' paints and varnishes, to speed up drying time. It is used as a wood preservative, in batik work, and is an ingredient in India ink. It is also a key component in linoleum and has the potential to be used for biodiesel.

Soap

The oil is used in soft soaps, shampoos, and hair tonics. It is reputed to restore hair loss.

Skin Lotion

Being rich in essential fatty acids, the oil moisturizes and protects the skin.

Luminant

The oil can be used as a luminant or used to make candles.

Tanning

The bark is useful in leather tanning.

Construction

Wood from the candlenut is used to make packing crates.

Dye

Fabric dye can be obtained from the roots and the nut husks.

Decorative

The polished nuts are added to leis that are worn around the neck.

Native Range:
Candlenut is native to Southeast Asia and Polynesia.

Aleurites Fordii
(Tung Oil Tree)

The tung oil tree is a rapidly growing tree or large shrub, 15 to 40 feet tall by 20 to 30 feet wide, with a flat top. The somewhat heart-shaped leaves can be up to 5 inches long. Its showy-white flowers have veins ranging from yellow to red and are 1 to 2 inches in diameter. When the tree becomes 4 to 10 years old, flowering is followed by fruit. The fruits are 2 to 3 inches in diameter and contain three to five oil-bearing seeds. They are covered by a thick hull that must be shucked before oil can be extracted from the seeds. Some commercial varieties produce very high yields.

Culture:
Tung oil trees require a region of high rainfall, mild winters, and fertile soil. In the United States they are generally limited to within 100 miles of the Gulf of Mexico shoreline. They need full sun and are somewhat finicky about soil quality. They prefer a moderately acidic, nitrogen-rich loam with fast drainage. They also favor mineralization with rock dust, such as colloidal phosphate. They are fairly drought tolerant and are probably at their best in Florida. They are cold hardy to zone 9 and are grown commercially in that region. Tung oil trees are natural hyperaccumulators of magnesium. Propagation is possible by sowing seeds in place or by transplanting seedlings when they are 1 foot tall.

Medicine:

The fruit extracts are antibacterial and might be beneficial to humans. It has been suggested as a natural remedy for bacterial diseases, but more research is needed.

Agricultural Uses:

Fertilizer
The oil cake byproduct is used as a nitrogen-rich fertilizer, but is poisonous if ingested.

Erosion Control
The tung oil tree can be used as a permanent topsoil stabilizer.

Other Uses:

Tung Oil
Commercial oil can be pressed from the tung oil tree nuts. Ripe fruits drop from the tree in the fall. The nuts are removed, and then dried and pressed for their oil. The oil content of the nuts is about 65%. Expect some varieties to contain more, while ordinary seedlings may have less.

Tung oil is much valued as a water and chemical resistant coating. It is used for waterproofing wood, paper, fabrics, electrical parts, gaskets, etc. It has been blended with other compounds to manufacture wall board.

Tung oil is also used to make paint and varnishes dry faster. Varnishes containing tung oil dry exceptionally hard and are very resilient. Tung oil is highly resistant to saltwater damage. Paints made from it can be used to coat boats and ships.

Fuel

The hulls of tung oil seeds have been used as a fuel source and could be a possible source of biodiesel. The resulting ash (potash) is used to make soap and glass.

Luminant

The oil from tung oil trees has been long used as an luminant in China, and the resulting soot (carbon) is used to produce Chinese ink.

Native Range:

The tung oil tree is native to western and central China.

ALLĪUM SATIVUM
(Garlic, Hu Suan, Ta Suan, Da Suan)
[Liliaceae Family]

Garlic is a perennial herb, often grown as an annual. The flat gray-green grass like leaves are about 1 foot long and up to 1 inch or more wide. The flower stems can reach 4 feet tall, and the plant spreads about 6 inches. The small white flowers form round clusters on top of the stems in early summer.

As a food and medicine, garlic is believed to have been in use since before recorded history. The Sumerians' use of garlic as a medicine is estimated to have begun about 6,000 years ago. The Egyptians first cultivated garlic between 3,500 and 3,200 B.C.E. By 3000 B.C.E. garlic was being cultivated throughout the Mediterranean region.

Garlic was introduced into China about 2,000 years ago. The first record of its use as a medicine in China was about 510 C.E. It is nothing short of phenomenal the way garlic was acclaimed in the written works of so many ancient cultures, including the Babylonians, Sumerians, Phoenicians, Egyptians, Hebrews, Greeks, Romans, Chinese, and Vikings.

Culture:
There are two types of true garlic, hard-neck and soft-neck. Hard-neck garlics are more cold hardy and better adapted to northern climates. They are generally more flavorful, and while their bulbs are smaller, the cloves are bigger and easier to peel. Hard-neck varieties are hardy to zone 4.

Allium Sativum

Soft-necks are the common commercial garlic found in the marketplace. Being less cold hardy, they are grown coast to coast throughout much of the south, but most commonly in the Southwest. In California they are grown in the south and also in the warmer areas of the north. Soft-necks typically have the highest yields and the largest bulbs, though the cloves are smaller. They have a milder flavor than long-necks, store better, and can be braided. They are also less labor intensive because they lack seed stems, which reduces the processing time.

Soft-necks seem to prefer slightly alkaline soil. Full sun and good drainage are important. Rich to moderately rich soil is preferred, with a pH 6.0-8.0. Garlic is drought tolerant. In a dry spring it is often irrigated to increase bulb size, however, withhold irrigation for a few weeks before harvesting when the leaves dry out.

Garlic is very easy to grow. Hard-necks, however, can be a bit more finicky than soft-necks. Nevertheless, it is hard to fail.* In climates where the soil freezes, plant garlic in early spring. Wherever frozen soil is borderline, try a heavy straw mulch with fall plantings. If the soil becomes very dry in winter, some added moisture can help impede frost penetration. As the temperature begins to rise in spring pull back the mulch on sunny days to let the soil warm up and then re-mulch.

---

*In the southern Willamette Valley where I live, I have watched a patch of weeds produce a dozen or more hard-neck garlic shoots year after year. They received no care of any kind, no irrigation despite the dry summers, and no weeding. The soil in this patch was liberally amended with toy soldiers, bicycle parts, broken glass, tin cans, and a variety of multi-colored bits and pieces of plastic.

Cloves are planted 6 inches apart and ½ to 1 inch deep with the pointed end up. In row plantings (as opposed to colonies), rows are spaced two feet apart. Colony plantings are designed for perennial yields; they are excellent for orchards and gardens.

To develop a perennial colony first prepare the site with compost and other appropriate amendments. Grade the planting surface. Although garlic is disease resistant, in poorly drained soil root rot is common. The planting area should be 3 to 4 feet in diameter. Space cloves about 8 to 9 inches apart, and then sit back and wait a few years—and keep it weeded. Once the colony is reproducing well, harvesting can begin. Always leave a few mature garlics scattered through the colony for mother plants. Any immature garlic bulbs you may pull up during the harvest should be replanted.

Fall plantings are harvested in early summer, and early spring plantings in mid to late summer. In most cases, annual harvesting begins once the tops have dried.

Dry the bulbs in a dry, shaded, well-ventilated location. When the dirty paper-thin outer layer dries, peel it back to the first clean layer. At that point they can go into storage. To store, garlic should be kept in a dry, dark place at room temperature. Brown paper bags suffice for darkness. If the air is too humid, garlic can sprout. If the temperature is too warm, garlic will degrade into a gray powder. Braided garlic can simply be hung from the rafters.

Varieties:

Hard-neck Varieties

Music: This variety has a moderately pungent flavor with a protracted aftertaste. An extremely productive garlic, it yielded over 13,500 pounds per acre in trials at Michigan State University. It has an exceptional storage life for a hard-neck—nine months to a year—and is also one of the easiest to peel.

Russian Red: The Russian Red's flavor is strong and hot with a nice sweet aftertaste. It has extra large bulbs for a hard-neck with six to eight cloves. Remarkably, this garlic tolerates soggy winter soils. It thrives in cool weather and is very cold tolerant. This Russian heirloom variety was brought to British Columbia in the 19th century by the Doukhobors, religious refugees from Russia who are noted as the first to settle much of Canada's interior.

Siberian: The flavor of the Siberian is fairly strong and considered excellent. It is another producer of extra large bulbs and is very productive as well. Its most notable feature is exceptionally high levels of the sulfur based compound allicin which is responsible for its pungency and many of its pharmacological actions.

Spanish Roja: Spanish Roja is considered to be gourmet quality with a rich spicy flavor. The bulbs are large for a hard-neck with about ten cloves per bulb and is excellent even eaten fresh. A good keeper too, it stores four to six months, peels easily, and is quite cold tolerant.

Soft-neck Varieties:

California Early: This white, commercial variety has a mild delicate flavor, produces large bulbs with ten to twenty cloves per bulb. It peels easily for a soft-neck.

California Late: The California Late variety has a strong high quality flavor. It is the longest keeping of all the white garlics, remaining fresh for at least one year. It is one of the best for braiding and for that reason alone it is quite popular.

Inchelium Red: This mild flavored variety has a delightful, long lasting aftertaste. It was a first place winner for flavor at a Rodale Food Center Program. The bulbs are exceptionally large, and it is very productive. It keeps for six to nine months, and the longer it is stored the hotter it gets. This variety is extremely vigorous. It was discovered on the Colville Reservation in northeast Washington State.

Susanville: Susanville's flavor is rated excellent. It is a very popular all-purpose garlic. Its bulbs are small to medium but have large cloves, a bit unusual for a soft-neck. It has exceptional keeping qualities, storing, in some cases, for well over a year. It is also a widely adapted variety.

Food:

Garlic is one of those uncommon herbs that is a veritable warehouse of diverse nutrients, so much so that garlic consumption is nearly synonymous with good health, in addition to being among the world's most popular condiments.

Although garlic's nutritional profile varies, like most edible plants, due to genetics, soil types, and other factors, garlic is often very rich in vitamin A, vitamin B1 (thiamine), vitamin B12 (riboflavin), vitamin B6 (pyridoxine), and vitamin C. It is also super rich in certain minerals and trace elements. Garlic contains over seventy-five sulfur compounds. This is uniquely important to garlic because when the cloves are cut, crushed, or chewed it converts to the potent sulfur compound allicin, while at the same time giving rise to others including ajoene, allyl sulfides, and vinyldithiins. Also abundantly present is the potent antioxidant selenium (garlic being the richest source known). Garlic is also rich in chromium, phosphorous, potassium, calcium, zinc, iron, copper, tin, and aluminum, plus 17 amino acids, and possibly more. James Duke in his book, *"The Green Pharmacy Guide to Healing Foods,"* says that this remarkable plant has over 2,000 biologically active elements that prompt its medicinal effects. Your health benefits considerably when garlic is a regular part of your diet. Duke has also listed 147 phytochemicals found in garlic.

Garlic also contains vitamin B3 (niacin), folic acid, iodine, magnesium, manganese, cobalt, germanium, and chlorine. It is about 6.2% protein, 30.8% carbohydrate, and 1.5% fiber.

It is hard to imagine another food as strongly flavored as garlic that is also so deliciously desirable and essential to the world's rich and varied cuisines. The cloves, flowers, seed, flower stalks, and the young tender leaves are all used. They may be baked, boiled, broiled, roasted, sautéed, and even used raw despite the repugnant odor inherited by its consumer.

Garlic's cloves can produce delicious garlicky sauces, or they can be used as indispensible condiments. The flowers too are used to prepare sauces. Seeds can be eaten as is or better yet sprouted. The flower stalks are used as condiments and the tender young leaves are great in soups or salads or as toppings for various recipes.

When we are cooking with oil, many oils go trans (bad) after just five minutes. But when combined with antioxidant rich foods like garlic, cooking oil can remain harmless for up to twenty minutes.

By the way, all you future moms should know that babies prefer the breast milk of moms who consume garlic to that of mothers who do not. That is according to two studies conducted on the subject.

Medicine:

Modern science seems to be confirming unequivocally that garlic's 6,000 years of medicinal use has some real validity behind it. More impressive, however, is the wide range of medical conditions garlic has been used to heal or mitigate in contemporary times.

Functions

Garlic is known as a potent antimicrobial, but it is further illuminated by its antibacterial, antiviral, antifungal, and parasiticidal actions. It is an excellent cardiovascular tonic that can serve as a hypotensive, vasodilator, hypocholesteremic, antithrombotic, and multi-functional cyclooxygenase inhibitor. It is also an immunostimulant, an anti-inflammatory, and an anti-cancer agent. It is hypoglycemic, diaphoretic, a febrifuge, carminative, analgesic, antiarthritic, antispasmodic, and alternative.

The most potent medical benefits are found in raw garlic. When raw garlic is cut or crushed a chemical transformation results. Alliin, a non-volatile odorless sulfur containing amino acid in garlic's bulb, comes in contact with the enzyme allinase, normally isolated from alliin in another part of the bulb. When the two interact, the alliin is converted into the pungent volatile sulfur compound allicin. Allicin, along with a handful of garlic's other sulfur containing compounds, is responsible for garlic's odor and a great deal of its pharmacological effects.

Just one-thousandth of 1% of allicin is sufficient to suppress the development of numerous microbial disease organisms. Unfortunately, allicin is unstable and can be destroyed by heat, as in cooking. Ten minutes of cooking destroys about 40% of its medicinal effectiveness.

Cooked garlic, however, can still be an effective medicinal

food. It will, for example, lower LDL cholesterol while raising HDL cholesterol. Cooked garlic helps prevent heart attacks and strokes by thinning the blood. It helps regulate the mucus membranes and acts as a decongestant. Cooked garlic provides relief from coughing and helps prevent bronchial disorders. A study of a group of garlic eating citizens in India by the National Cancer Institute concluded that those who eat large quantities of garlic and other Alliums were nearly twelve times less likely to get gastric cancer than those who ate little or no Alliums. Research also suggests a similar protection against colon cancer.

Garlic is so effective in healing or controlling such a wide range of medical conditions it would require an entire book to thoroughly document them—and is well beyond the scope of this profile. For just a sampling, it is known to control seventy-two different infectious diseases. In one or another of its antimicrobial forms it is used to treat colds, the flu, some coughs, including chronic bronchitis and whooping cough, pneumonia, diphtheria, amoebic or bacterial dysentery, diarrhea, ring worm, pulmonary tuberculosis, typhoid, botulism, Candida albican, yeast infections, vaginitis, streptococcus, trichomonads, sinusitis, internal parasites, human immunodeficiency virus (HIV), etc. A single milligram of allicin, garlic's very potent sulfur containing compound, is reputed to be equal in its antibacterial effectiveness to fifteen standard units of penicillin.

Circulatory System
Garlic taken as a tonic tones and strengthens the heart and protects it in a variety of ways. Dozens of studies have demonstrated garlic's antihyperlipidemic action of lowering elevated LDL cholesterol and serum triglycerides, both of which can lead to heart attacks as well as strokes if not

normalized. Not only can garlic reduce LDL cholesterol as much as 12%, but just a single clove of raw garlic consumed daily has been shown to reduce LDL by 9% on average. Indeed, garlic has been shown to be even more effective at lowering LDL than the pharmaceutical drug clofibrate.

In the case of triglycerides, a few dozen studies, most of them double blind, using garlic pills that contained 600 to 900 milligrams of available allicin, reduced abnormally high triglycerides by 8 to 27% when taken daily for one to four months. More on these pills later.

Garlic also helps prevent the build up of fat deposits on the inner linings of arteries that can constrict blood circulation and lead to the disease known as atherosclerosis, which is a known cause of heart attacks and strokes. This action has been confirmed in both human and animal studies. Additionally, it has been shown that just two of garlic's constituents, the sulfur containing allicin and ajoene, safely and effectively help prevent blood platelet aggregation which can cause blood clots and lead to heart attacks or strokes.

In the case of hypertension, garlic helps normalize high blood pressure. Numerous human double-blind studies have shown that a daily dose of 600 to 900 milligrams of raw garlic lowers abnormally high blood pressure effectively, rapidly, and in a protracted manner. Garlic is a natural blood thinner, due to the action of methyl allyl trisulfide, one of garlic's active sulfur compounds. Garlic also helps expand the walls of blood vessels and promotes good circulation.

Anticarcinogenic

The National Cancer Institute has rated garlic's sulfur compounds near the top of their list of natural anticarcinogenic substances. One way garlic inhibits the formation of carcinogenic compounds is by stimulating the powerful "natural killer" (NK) T-cells that destroy cancer cells. Another way, also borne out by considerable epidemiological research, is that garlic and its constituent allicin block the proliferation of cancer cells by promoting apoptosis which inhibits tumor growth and determines the life span of cells. One study done by researchers at Akbar Hospital and Research Center in Panama City, Florida found that the NK T-cells of those who ate raw garlic or certain kinds of garlic extract destroyed 159% more tumor cells than the NK T-cells of those who consumed no garlic. In other human studies, it was found that garlic can suppress the formation of nitrosamines, potent cancer causing compounds that develop during the digestive process.

Research has determined that garlic extract may inhibit the growth of skin cancers such as melanoma by 50%. Human population studies have also shown that eating garlic on a regular basis reduces the risk of colon, esophageal, and stomach cancers. Epidemiological evidence indicates that in addition to cancers of the gastrointestinal tract, garlic can suppress other kinds of cancer as well, such as cancerous tumors of the breasts and skin. Japanese scientists found that when breast cancer was induced in mice, garlic completely eradicated it. Experiments in the United States revealed that garlic is more effective for treating bladder cancer than some cancer vaccines. In China researchers found that garlic significantly reduced the risk of prostrate cancer.

Antioxidants
Garlic's antioxidants are critical to many of its disease preventing and healing actions. Two very important flavonoids found in garlic with powerful antioxidant actions, quercetin and rutin, do much to improve the immune system, as do garlic's vitamins beta carotene, B3, C, and E. Garlic is

also an excellent source of such antioxidants as manganese, selenium, and zinc.

One of the ways many of these antioxidants contribute to a strong, resilient immune system is by producing T-lymphocytes, a type of white blood cell known as T-cells. There are three kinds of T-cells, the helper T-cells, the suppressor T-cells, and the NK T-cells. The before mentioned NK cells are the heroic warriors of the immune system and its first line of defense against disease. NK cells roam the circulatory system seeking aberrant cells that have been damaged, become diseased, or cancerous. Once the NK cellular defense force finds these aberrant cells, they are spontaneously and aggressively obliterated with NK's chemical weapons. At the same time, NK cells send out chemical messages requesting the support of specialists like macrophages which destroy worn out cells and cellular debris. When our bodies contain sufficient numbers of NK cells not even cancer or AIDS can survive them.

Respiratory System
Garlic also offers significant benefits for the respiratory system, especially the upper respiratory system. Its healing functions are anti-inflammatory and anti-microbial, plus it can help to expel excessive mucus from the lungs. It serves as both a preventative and treatment for asthma, bronchitis, catarrh, colds, hay fever, influenza, and whooping cough.

Blood Sugar Benefits
The sulfur containing compounds allicin and allyl propyl disulfide receive most of the credit for garlic's hypoglycemic action. Garlic is most effective through an interaction with insulin, and it helps the pancreas secrete insulin. In addition, garlic also helps thwart cardiovascular problems that have

been linked to diabetes mellitus, rendering Non-Insulin Dependent Diabetes Mellitus (NIDDM), which associated with pancreatic diseases such as chronic or recurrent pancreatitis, a secondary condition. Studies performed by scientists from the West and from Asia, particularly China, have demonstrated garlic's ability to help prevent and treat Type 2 Diabetes and NIDDM.

Gastro-intestinal System

Conditions of the gastro-intestinal system can benefit from garlic treatment. In traditional Chinese medicine (TCM) garlic is used to treat enteritis—the inflammation of the mucosa and submucosa of the small intestines. Garlic can also provide some relief from irritable bowel syndrome due to its prebiotic influence that promotes the production of beneficial intestinal bacteria. Research in India has shown that garlic's prebiotic effect benefits the digestive tract, stimulates good digestion, and helps eliminate diarrhea, as well as increasing the body's ability to absorb minerals.

Antimicrobial

Ear infections caused by pathogenic microbes (viral, bacterial, fungal) respond to treatment with garlic oil's antimicrobial and anti-inflammatory actions. For example, acute otitis media (AOM) is an infection of the middle ear that produces fluids which can block the ear canal and cause inflammation. Provided the eardrum is not perforated, herbal eardrops made with garlic in an olive oil base can relieve this condition. Some practitioners of herbal medicine may add other herbs.

Nervous System

Garlic also provides relief from nervous system disorders. It is an antinoceptive that helps control nociception, a painful condition where specific nerve fibers are stimulated and broadcast irritation to the central nervous system's peripheral nerves. Headaches, including migraines, anxiety, and insomnia are other nervous system conditions that may benefit from garlic.

Miscellaneous

Contemporary Chinese pharmaceutical journals have reported that garlic and its various preparations have successfully treated acute appendicitis, hypersensitive teeth, infectious hepatitis, and trachoma. In one study, eleven patients infected with the often fatal cryptococcal meningitis were treated with garlic. After several weeks of therapy, all eleven recovered from the disease.

In TCM garlic is used to heal carbuncles, diphtheria, insect and snake bites, nose bleeds, and sores. In addition the Chinese use garlic in the treatment of many of the same conditions it is used for in the West.

Other ills garlic has been used to treat in the West include fatigue, fever, insect bites and stings, Raynaud's disease, and skin conditions.

Dosage

To benefit from garlic's rich source of preventative or therapeutic properties at least 4,000 milligrams of fresh raw garlic, or its equivalent in garlic products, is required. Four thousand milligrams amounts to one to four cloves of raw garlic daily. As many as three to five cloves are common recommendations to treat various ills.

Garlic has been made into commercial products, typically in pill form. To be effective, which some are not, they must contain 4,000 milligrams of raw garlic powder. Murray and Pizzorno report in their book *Encyclopedia of Natural Medicine* that numerous clinical studies have recommended that commercial garlic products should be standardized at 10 milligrams of allicin.

Murray and Pizzorno also point out that while allicin is the most beneficial of garlic's compounds, there are others like S-allylcysteines and gamma-glutamylpeptides that are medically significant as well. Thus the ideal commercial product should contain as many of raw garlic's beneficial constituents as possible.

When garlic pills receive an enteric coating, they are able to pass through the stomach without the allicin being destroyed by stomach acids. Once it has safely made its way into the small intestine undigested, it completely dissolves into the bloodstream.

Another benefit these products offer is that they prevent garlic's odor from escaping from your pores and offending those around you.

Warnings and Considerations:

Individuals that have sensitivities to garlic may need to avoid high doses. Garlic can act as an anti-inflammatory for some gastrointestinal conditions, but for people with highly sensitive systems, it might further inflame the problem instead. Garlic can be used to treat gastric ulcers, but it can actually make them worse in some individuals.

In high doses, garlic can irritate the intestinal mucosa resulting in diarrhea, nausea, and vomiting. Heartburn and gas are also common.

Large doses of garlic should be avoided it you are taking a pharmaceutical anticoagulant or an antihypertensive drug. Check with a holistic physician first.

Skin inflammations can occur with protracted handling, in farm work for example.

Do not grow garlic on selenium rich soils. It will become toxic.

Driving people away because you smell of garlic may not be a safety issue, but it can be a personal issue. People who want to avoid becoming a social outcast often combine raw garlic with chlorophyll supplements or chlorophyll rich foods like nettles and parsley to mask the garlic odor. This is fairly effective although do not expect to be totally free of the smell.

Agricultural Uses:

Insecticidal

The Henry Doubleday Research Association in England studied garlic's insecticidal potential for eight years. They concluded that using emulsified garlic oil as an insecticide is as effective as "commercial organo-phosphorous or organo-chlorine insecticides." Yet the garlic oil is non-toxic to people and animals even in high concentrations, nor does it persist in the environment once it has done its job. Researchers had an 89% kill rate with aphids and a 95% kill rate of onion flies.

Over time farmers and gardeners using homemade garlic insecticides and garlic-based feeding deterrents have reported successfully controlling or repelling caterpillars, codling moth, grasshoppers, grubs, leafhoppers, scale, thrips, aphids, Colorado potato beetle larva, nematodes, and white flies. The Chinese may have been the first to recognize the value of garlic's insecticidal properties, and they have long used it to control aphids and mites. Just intercropping garlic with onions has reduced damage from onion flies. Other field trials demonstrated that a 3% garlic oil emulsion spray repelled black pea aphids and other insect pests for close to thirty days.

Although emulsified garlic oil is a potent, biodegradable, nontoxic insecticide, it does not discriminate between insects that are pests and insects that are beneficial. For instance, it is fatal to lacewings, lady bugs, and syrphid flies—all important allies in pest control. For this reason, it should only be used as a last resort.

Other research reports on the benefit of garlic for the protection of stored grains. This is accomplished by placing garlic cloves on top of the grain.

Antimicrobial

Garlic's antimicrobial action is reported to help cure various diseases affecting stone fruits, nut trees, and vegetables like tomatoes, potatoes, cucumbers, radishes, spinach, and beans. Research at the University of California demonstrated garlic's antimicrobial power against angular leafspot, anthracnose, bacterial blight, downy mildew, early blight, rust, and scab.

Wildlife Repellant

Garlic repellants that are available in the marketplace today can protect orchards and other plants from deer predation. At least some of these repellants remain effective for years without harming the trees.

Guild Associates

As companion plants for fruit trees, garlic is believed to repel borers. When grown with roses or raspberries, garlic is reputed to keep them free from attack by Japanese beetles.

Livestock

One would assume that when animals are affected by the same ills as humans they would benefit from the same garlic treatments that work on humans. For instance, Russian scientists used garlic extracts to treat thirty rabbits suffering from infected wounds. The result was an increased rate of healing. In another case, animals infested with ticks carrying encephalitis, a situation that can result in fatalities, were treated with volatile fractions of aqueous and alcoholic garlic extracts. The ticks stopped increasing in size and number in just twenty minutes. Ten minutes later they dried up. From that point on, the garlic repelled new infestations. In the case of cattle suffering from hoof and mouth disease, they were also healed by garlic. The Henry Doubleday Research Association found that when chickens and rabbits were fed garlic their overall health improved. Sheep and goats receive beneficial nutrients when they are fed dried garlic leaves and stalks.

Other Uses:
Public Health
In India, garlic's insecticidal action has long been recognized and used to kill mosquito larvae, houseflies, and other insect pests. Entomologists at the University of California sprayed garlic oil, at a mere 200 parts per million, on mosquito breeding ponds and had a 100% kill rate for five different species of mosquito larvae. Even species highly resistant to insecticides perished.

Glue
In his book *Plants for a Future,* Ken Fern informs us that garlic makes outstanding glue when its juices are used to repair glass and ceramic items.

Native Range:
Garlic's native origin is believed to be the upland plains of West-Central Asia, but it seems to be extinct in the wild.

ASCLEPIAS GENUS
(Milkweeds)
[Apocynaceae Family – Asclepiadaceae Subfamily]

There are approximately 200 species of Asclepias. Most are found in North America and Africa. Although nearly half the species grow in North America, only two of the most remarkable species are profiled here—Asclepias speciosa or "showy milkweed," which is native to much of the western half of the continental United States, and Asclepias syriaca or "common milkweed," which is native to most of the eastern continental United States. Their ranges overlap in the middle of the country where they often hybridize. Not only do these two species look very much alike, they have a great deal else in common as well.

Asclepias Speciosa
(Showy Milkweed, Silkweed, Lechero, Pink Milkweed, Lechones)

Showy milkweed is a distinctive perennial herb (see cover photo). It is a stocky upright plant, coarse but nonetheless attractive, particularly in flower. Sometimes the rhizomes creep and colonies form, although only occasionally invasively. Typically showy milkweeds are 2 to 4 feet tall. On a harsh site they mature at little more than 1 foot; however, in a favorable habitat and with good genes, they might reach 5

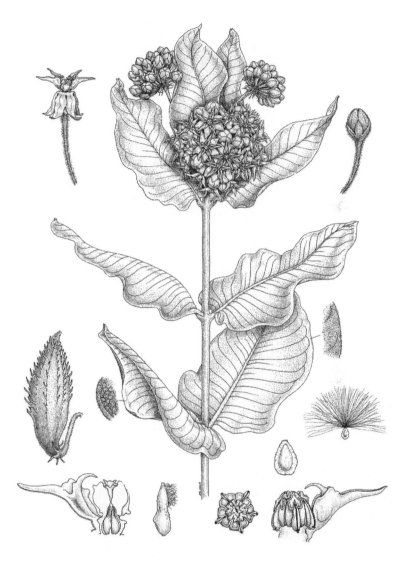

Asclepias Speciosa

or even 6 feet tall. The bold, oval or oblong leaves are broad with distinct veins and are arranged opposite each other on the stem. Leaf sizes range from 4 to 10 inches long and 1 to 5 ½ inches wide.

All parts of the plant contain a milky sap or latex, but it is particularly evident in the leaves and stems when they are cut or wounded. The light pink to rose-purple flower clusters are showy and delightfully fragrant. Each small flower presents a 5-pointed star face. Numerous flowers grow together in dense spherical umbels (clusters), 2 to 3 inches in diameter. Depending on habitat, altitude, and latitude, flowers might appear in May and continue into September. When fully grown, the coarse spindle-shaped seed pods are about 3 to 5 inches long. The pods dry as they mature and split open to allow the numerous seeds they contain to escape. Each seed has a silky tail that serves as a sail to carry it in the wind. The seeds are dark brown while their feathery tails are gleaming white.

Asclepias Syriaca
(Common Milkweed, Virginia Silk, Silkweed, Algodoncillo, Herbe a la Ouate)

A deciduous perennial herb, common milkweed is a single stout stalk, usually 2 to 5 feet tall but occasionally up to 6 ½ feet tall. It has broad, distinctive, elliptical opposite leaves with well defined veins. The leaves are 3 to 12 inches long and up to 4 inches wide, borne on small stems off the single stalk. Its flowers form into spherical clusters and their colors vary, sometimes pinkish or rose purple or greenish purple with some highlights. They bloom from late May or early June to August and are often fragrant. The fleshy gray-green seed pods are spindle-shaped, warty, and about 5 inches long. They are borne solitary or in pairs and turn brown before they split open and release the silky-tailed seeds inside. While this distinctive plant is sometimes described as coarse, it might be more accurate to say it is uniquely beautiful.

All parts of the plant contain a milky sap rich in hydrocarbons. Although common milkweed and showy milkweed are very unique looking compared to other plants, they are not easy to distinguish from each other.

Culture:

Milkweeds want full sun and will grow on almost any soil, but good drainage is essential. They are most common on dry loams or sandy and gravely soils, but tolerate calcareous soils. Common milkweed does well with well-drained clay soils. Although quite tolerant of drought, it performs better with additional moisture. In the wild milkweed often seeks out stream banks, depressions, and swales where moisture tends to be more concentrated, and is typically found at low to mid elevations. Like most perennials, milkweeds first flower and produce fruit in the second year of life. Considered weeds because they might be found among other crops, showy and common milkweed are well-suited to marginal lands in that they are drought resistant and tolerant of high winds, poor soils, and other difficult conditions. They are hardy in zones 4 to 8.

In the case of biofuels, milkweed can thrive on marginal land too poor for most crops. This is a sharp contrast to corn that demands the best soils and substantial water to mature satisfactorily for ethanol production. In addition, milkweed is a perennial that can be harvested repeatedly for years without replanting. This saves both labor and top soil. On the other hand, when milkweed is cultivated on a large scale as a mono-crop, it, like other monocultured crops, develops serious problems. Bacterial blight and black spot fungus are the most damaging.

Milkweed propagates easily from its highly viable seed. Collect the seeds from ripe pods, just before the pods split open if possible. Sow seeds directly in the fall. When sowing seeds in early spring, they should first receive three months of cold stratification.

Propagation by root division is also reliable. Roots should be dug and divided only while dormant, and each divided piece requires at least one bud. For best results, irrigate milkweeds during their first year and keep them weeded. Burn the dead stalks and flowers and seed production will increase. Burning stimulates new stems to grow taller and straighter. This is of particular benefit when stem fibers are to be utilized for developing useful products.

Stem fibers are harvested in the fall after the foliage starts to shed and stalks turn tan or gray. At this point the stems should be good and dry, and if they can be snapped off easily at ground level, they are ready. This will encourage the plants to sprout in the spring.

Food:

Showy milkweed and common milkweed are quite edible and to many palates delicious. However, without proper preparation they are bitter and toxic. These plants should not be eaten raw, but when parboiled for one minute in two or more changes of water, they become safe and desirable. This treatment removes the toxic cardiac glycosides as well as the undesirable bitterness. After each change of water, they must be immediately transferred into boiling water. If not they will retain their bitterness. Once you have finished the process of changing the water twice, let the vegetables boil at least ten minutes before eating.

Throughout the growing season and after, these milkweeds provide a sufficient bulk of foodstuffs to have made them economically important to First Nation peoples' diets in regions where milkweed is found. The new shoots, succulent young leaves and stems, unopened flower buds, the flower clusters, immature fruit pods, seeds, and the root have all been used for food. The Spanish colonists and other European settlers wasted no time integrating milkweeds into their diet. Today, milkweeds are a favorite among many wild food enthusiasts.

Sprouts

Early in the season the newly emerging succulent sprouts, reputed to be high in vitamin C, are the first fare that these two milkweeds yield. The tender shoots are gathered while they still snap off easily at the ground, rather than bend, or when about 4 to 8 inches tall. As with all the vegetative parts of this plant, the sprouts must be parboiled to expel the toxins and the bitter taste. Bring the water to a boil then add the plant material. Boil for about a minute then change the water. Repeat this process two or three times. Move the sprouts directly to pre-boiling water each time. Boil the fresh water first before adding plant material.

After parboiling, they should be boiled until tender. Usually, this takes ten to twenty minutes. Recipes for asparagus work well for these shoots. In modern times they are seasoned as asparagus might be. Some palates prefer them to asparagus.

Leaves

Once past the sprout stage, milkweed's tender tops, small succulent leaves, and stems can be pre-treated then boiled until tender—about as long as the sprouts or a little longer.

A few California tribes were known to use the herbaceous parts of showy milkweed as a thickener by mixing it, for example, with their manzanita-berry cider.

Flowers
The flower clusters in the bud stage are uniquely delicious. After parboiling, they are boiled an additional ten or twelve minutes until tender. Add them to soups and meat dishes as was the practice of native people East and West.

Once the nectar-rich flowers open, the clusters can be boiled down to a thick, sweet syrup, somewhat reminiscent of brown sugar. The dew that collects overnight in the flowers may be collected and boiled also to make a similar sugary syrup. This latter approach saves the flowers for beneficial insect forage and allows the seedpod stage to develop. Some tribes used the dew found in the showy milkweed's flower clusters to sweeten wild strawberries. Dew, after all, is the perfect medium to extract the flower's sweet nectar.

Seedpods
The immature seedpods provide the next tasty vegetable dish. The young pods are collected when they are about 1 to 1½ inches long, a job less tedious than hand harvesting beans. After pre-treatment, boil them for ten to twenty minutes until tender. Pueblo peoples and other tribes north and east cooked the young pods with meat stews because they would tenderize the meat while adding flavor and bulk to the stew. (A substance found in these milkweeds called asclepain is an effective substitute for papain for tenderizing meat and rendering it more digestible.) The young pods also make a welcome addition to cooked grains. Some people favor the Cheyenne preference of peeling the pods before eating.

The seeds can be singed to remove their silky tail, then

dried and ground into flour. The seeds have also been sprouted and eaten. A few tribes even boiled and ate the roots. Once properly prepared the roots were usually added to meat dishes.

Sap

Several tribes were fond of using the plant's milky latex as chewing gum. Despite the sap's cardiac glycoside content, it appears they suffered no ill effects from chewing it. They would collect the latex by breaking the leaves along the mid rib or snapping the leaf off from the stem causing the sap to flow freely from the wound. They would then harvest the latex and heat it slowly, stirring it all the while, until it would solidify. Most tribes added some kind of fat, like deer or salmon fat, to the latex as a binder; otherwise it quickly disintegrates into small bits when chewed.

Medicine:

Showy milkweed was an important First Nation remedy for numerous conditions. It was used as a cathartic, an expectorant (to loosen and expel excess mucous), a diuretic, and in stronger doses, as an emetic (causing the user to vomit). There is no doubt that consuming a sufficient amount will induce vomiting if the plant is uncooked.

The constituents in showy and common milkweeds closely resemble those found in all Asclepias species. The toxic principle (cardiac glycoside), for example, is in all the species. However, it is more abundant in Asclepias species with narrow leaves than in those with broad leaves, like the two species in this profile. Collectively the Asclepias contain other kinds of glycosides, resinoids, and tiny amounts of alkaloids, all of which are present throughout each plant. It is these compounds that are believed to account for the healing influences of these plants. The latex sap possesses asclepain, an active proteolytic enzyme that may add to some of the other medicinal effects.

Latex Sap

Showy milkweed is 75% latex, and its latex is found throughout the plant. The latex sap of common milkweed was used for medicine by the Catawbas, the Cherokee, the Iroquois, the Ojibwas, and the Rappahannock nations. They made a salve from the milky latex and applied it daily to remove warts, moles, corns, and calluses. This same ointment was used as an antiseptic and to treat sores, cuts, wounds, burns, ringworm, syphilitic eruptions, and bee stings.

It has also been used in more recent times as an antiseptic in pharmaceutical medicines. Medical research of this floral genus, however, is scant. A few laboratory tests using other Asclepias species have hinted at their as yet untapped potential in modern medicine.

In one such lab test, an extract of amplexoside glycoside inhibited the growth of cancer cells. This is of particular interest since Asclepias has been used in folk medicine as a cure for cancer. In animal tests, one species demonstrated an antibiotic action on a particular strain of tuberculosis, while other studies conducted on various animal species had a stimulating action on the uterus.

Roots

The Flathead peoples of the Montana region used the roots of showy silkweed as a remedy for indigestion (as do some herbalist today). They called it "stomach ache medicine." The roots were cut up, dried, and crushed, then boiled and taken as a tea. Sometimes they simply chewed on a piece of fresh root.

The Chippewa used the roots of common milkweed in a cold decoction that was consumed with food to increase postpartum milk flow. This use may have been inspired by milkweed's milky sap, as in the doctrine of signatures. For chest distress, the Menominee used common milkweed as a root decoction or simply ate the flower buds. The Cherokee used a root infusion to treat venereal disease and to treat inflammation of the breasts (mastitis). It was also used as a laxative. The Cherokee also use common milkweed in the treatment of edema and kidney stones. (Some herbalists today still use it to treat chronic kidney conditions.) Other tribes prescribed a root decoction for asthma, dropsy, and dysentery.

Sometimes a poultice from the crushed root was applied as a topical for sprains, rheumatic swelling, and other swollen parts. A decoction prepared from the boiled root was used to remedy asthmatic conditions, measles (also boiled leaves), and gonorrhea. A contraceptive tea prepared from the boiled root was taken to induce temporary sterility. A tea was also made from the roots that was used to treat measles and coughs.

Today milkweed root might be prescribed as a diaphoretic, a diuretic, or an expectorant. The roots are dug in late fall after the plant becomes dormant. The fresh roots are cleaned and chopped into small pieces, then dried and stored for use when needed.

Leaves

Some western tribes ingested a tea prepared from the leaves of showy milkweed for colds and coughs. Native women recovering from childbirth drank a tonic prepared from the entire plant. Nursing mothers took a cold infusion of the whole plant to improve lactation. For sore or tender breasts, they used the same infusion.

A compress of the plant, after it had been boiled and then cooled sufficiently, was plastered on the head to relieve headaches. An eye medicine for temporary blindness, such as snow blindness, was prepared by boiling and straining the vegetative tips of the plant. A clean cloth was used to apply this decoction to the eyes.

The Iroquois used an infusion of common milkweed leaves for stomach difficulties. To prevent hemorrhages at childbirth, they used an infusion of different parts of the plant mixed together.

The Meskwaki and Mohawk nations used common milkweed as a contraceptive. The Mohawk dried, crushed, and boiled the above-ground plant and mixed in Arisaema spp.— Jack in the Pulpit. (Be forewarned, consuming Jack in the Pulpit's corms raw or cooked can cause an intense burning sensation in the mouth due to the presence of calcium oxalate crystals. Only thorough drying can eliminate the effect.)

Seeds

At least a few western tribes relied on showy silkweed to treat rattlesnake bites. First the wound was excised and the venom sucked out as much as possible. The boiled seeds were then packed on the bite to extract more venom. A variation of this treatment was to apply the plant's sap. A salve was also prepared from the crushed seeds after their silky tails were burned off. This salve was then applied to sores and scrapes to soothe and heal an injury.

Considerations and Warnings:

These two broadleafs, showy and common milkweed, contain toxic levels of cardiac glycosides. This is one reason it is essential to prepare these plants properly before ingestion. Although cardiac glycosides are very good for the heart in very low doses, in higher doses they become poisonous. The line between the two is uncomfortably close.

Agricultural Uses:

Beneficial Insects

Milkweed flowers attract beneficial parasitic wasps and flies that can control common garden and orchard pests. For example, the flowers play host to the predators or parasites of codling moths, cabbage worms, and tent caterpillars making them ideal for inter-cropping with other plantings.

Forage for Honeybees and Other Pollinators

The heavy nectar of showy and common milkweed flowers attracts honeybees. In locales where milkweed is abundant, a bee colony might produce 60 to 100 pounds of honey from this plant alone. The honey has a light color and a pleasant flavor.

The flower's sticky pollina represents some danger to the would-be pollinator. If the insect is not strong enough to pull itself free, its life's journey will terminate at this spot. When struggling to break loose from the pollina, a bee may sacrifice a leg to the effort. The pollen of milkweed is heavy and requires strength to transport. Occasionally a careless bee may become so weighted down by the pollen it becomes too heavy to fly.

Insect Repellent

An extract of the sap of showy and common milkweeds acts as a feeding deterrent. Seeds commonly vulnerable to wireworm predation become repellent to wireworms when coated with milkweed sap. When the sap is applied directly to soils infested with plant pests, they may also be repelled. Their chemical residues, however, may be allelopathic (toxic) to some crops. On the other hand, they have long been thought by many traditional gardeners to repel aphids from tomatoes simply by proximity.

Veterinary Herb

Horse sores and wounds have been treated with the latex sap. Horses, sheep, and cattle avoid eating this plant probably for the bitterness. Nonetheless, they are sometimes poisoned when provided with cut and baled hay from fields infested with Asclepias species.

Fodder from Fuel Processing Waste

Preliminary investigation of the residual plant material rendered from processing showy milkweed suggests that it is non-toxic and as easy to digest for sheep as alfalfa.

Other Uses:

Bio-crude Oil

Researchers first used the latex sap of common milkweed to produce rubber during World War II. However this proved difficult and expensive at the time, and it was subsequently abandoned.

The oil crisis of the 1970s sparked new interest in Asclepias' milky latex sap. Scientists searching for potential sources of synthetic fuels, primarily in plant saps, found Asclepias latex satisfactory as a crude oil substitute or as a useful synthetic rubber. Research at the United States Department of Agriculture (USDA) with Asclepias latex sap produced heating and fuel oil as well as plastic products. After the latex was separated from the plant, the unused vegetable by-products were processed into pellets that were then burned to generate electricity.

Standard Oil of Ohio was the only oil company to explore

milkweed's potential as a petroleum substitute. After five years of cultivating and processing common milkweed, Standard Oil did an analysis of their data and concluded that the cost of growing milkweed was too high and that the yields of oil were too low to be economically feasible when compared to the cost of petroleum at that time.

Not long after we entered the new century, renewable fuels from plant biomass generated scientific inquiry again. Processed showy milkweed yielded 58% liquid fuels and 11% light gases, making them very productive as a substitute for gasoline, although much less so for diesel and heating oil. Common milkweed was also very productive, generating 51 to 57% liquid fuels and 15 to 21% gases by weight.

Both of these species can provide a renewable source of bio-crude that could play a significant role in this era of climate change. Both species are adapted to marginal lands and a wide range of climates. In this country, the amount of land unsuitable for agriculture is enormous. This means milkweeds and other biofuel crops adapted to marginal conditions do not need to compete with fruits and vegetables for farmland the way ethanol does now.

While milkweeds have few problems with pests or disease in natural settings, from an agricultural perspective, monocropping exposes them to serious pest predation. For this reason, polyculture systems combining marginal land crops with milkweeds are more appropriate. The cultural requirements of the other crops should overlap with those of the milkweeds', and care in planning would be required to eliminate or minimize competition when necessary.

Insulated Fill Material

First Nation tribes have long used the silky seed tails of milkweeds for a variety of purposes. For hundreds and possibly thousands of years, native people stuffed mattresses, pillows, comforters, even hats with the seed down. They used it to line baby cradles to make them soft, warm, and very cozy. They even wove them into silky linens, a use that the French adopted in the first half of the seventeenth century. During World War II, the U.S. Navy used milkweed down to fill life preservers for its buoyant quality.

In 1987, following Standard Oil's termination of research of milkweed as a petroleum substitute, Herbert Knudsen, who had been the manager of Standard Oil's corporate ventures, decided to purchase the project and founded the Natural Fibers Corporation. Knudsen envisioned a new agricultural industry that used milkweed's silky seeds as a substitute for goose down.

The idea was to use the silky seed tails (floss) to stuff coats, vests, sleeping bags, comforters, and pillows. Knudsen brought in the Kimberly-Clark company to assist in research. The initial research indicated that milkweed seed tails were superior to down in a number of ways.

Here the tale gets murky. The Department of Textiles, Clothing, and Design at the University of Nebraska-Lincoln and the Institute for Environmental Research at Kansas State University-Manhattan worked together on milkweed research. Their results were published in the *Textile Research Journal* (60, #4, April 1991) in an article titled "Evaluation of Milkweed Floss as an Insulative Material."

Their evaluation concluded that down was superior in loft and compressibility to milkweed floss and that milkweed floss could not be recommended as a standalone fill material. When it was blended with 50% down, however, quality products could be manufactured that compared well to 100% down. Such blends make sense economically because milkweed floss sells for about half as much as down does. If its demand and production increases, the price may be reduced further.

The researchers also found that the insulative value of down and milkweed floss is nearly the same until the product (coat, pillow, etc.) is cleaned. Cleaning reduces milkweed floss' insulative value, and as fill it becomes matted and lumpy, while down remains nearly unchanged.

But the story wasn't quite finished. The 1992 USDA Yearbook of Agriculture titled *New Crops, New Uses, New Markets*, subtitled *Industrial and Commercial Products from U.S. Agriculture*, had a chapter called "Milkweed: The Worth of a Seed" by Renee Y. Sayler, Associate Director of Industry Development, University of Nebraska-Lincoln and Herbert D. Knudsen, President of Natural Fibers Corporation of Ogallala, Nebraska. They found some glaring differences between their research and that of the previous "Evaluation of Milkweed Floss as an Insulative Fill Material." For instance, they reported that milkweed floss had 25% more insulative power than down and that it was actually 20% warmer by unit of weight. In addition, they stated that milkweed floss is non-allergenic, has a fill power similar to high-grade goose down, and is more durable and breathable as well, even when it has absorbed moisture. Nonetheless, they too favored the idea of combining the two.

As an agricultural crop, milkweed cultivation requires conventional farm equipment with only minor modification. The same is true of processing equipment. Experimental plots of milkweed show yields of a little over 250 pounds of seed down per acre. Before the down is released, the green seed pods that are produced in the plant's second year are harvested with a modified uni-system corn picker. Generally, ten pounds of dried seedpods yield two pounds of down, three pounds of seed, and the remaining five pounds are pod husks.

Today most farmers and ranchers view milkweed as a noxious weed. In the future, however, this remarkably useful plant may become an invaluable component in an ecologically sustainable paradigm.

Byproducts from Milkweed Floss Production

Fireplace logs and cat litter have been made from seedpod biomass, while the bast fiber has been used to produce paper. It has also been suggested that milkweed floss could be blended with cotton for weaving into a linen substitute.

Cosmetics

Oil pressed from the seed has been evaluated for potential use in cosmetics or as a lubricant.

Fiber

The fibers found in the bark of milkweed stems are strong, glossy white threads, comparable to hemp or flax. Various tribes used these fine, very long fibers to make coarse cloth, rope and cordage, fish lines, nets, slings, and snares. This fiber was also important in the manufacture of native basketry. Milkweed fibers swell when wet helping to waterproof the baskets. The antiquity of this "use" by early tribal groups is suggested by baskets found in Utah's Danger Cave. According to radiocarbon dating, Danger Cave was first inhabited over ten thousand years ago.

Stems for fiber were collected in late fall or early winter, about the same time the roots were harvested for medicine. The fiber was liberated from the stems by vigorously rubbing and pulling the stems back and forth across a solid object like a block of wood or a rock. Another smaller rock was sometimes used to pound the stems until the fibers were freed from the stem's hard core.

Erosion Control

The deep thick roots of common and showy milkweed and their tendency to create colonies make it a useful soil binder. Their shallower feeder roots can spread 15 feet and can also help anchor the soil.

Ecological Functions:

Insectary Plants

Taken together, both showy and common milkweed host a surprising number of creatures. Monarch and viceroy butterflies receive protection from predators from the milkweed's toxic chemicals. Wild bees and other beneficial insects are served by the flower's nectar. True bugs like common milkweed bug and the large milkweed bug, plus various beetles like the red milkweed beetle and blue milkweed beetle, are also supported by these milkweeds, as are various moths.

Although milkweeds have potent defenses against predation, all of these insects have developed a certain amount of adaptation to milkweed's toxic chemicals. Although hummingbirds are definitely not insects, they are consumers of milkweed's nectar as well.

Symbiosis

Monarch butterflies (Danaus plexippus) have a uniquely intimate relationship with common and showy milkweeds. The larvae of the monarch, a green and yellow caterpillar, will feed only on the leaves of these plants. While this renders the monarch utterly dependent on the plant as a source of food for its larvae, it provides the monarch with an invaluable defense against predation. The milky latex is quite toxic to the butterfly's predators. The monarch's larvae, on the other hand, are immune to the sap's debilitating effects. The larvae consume sufficient amounts of the sap to endow the emerging butterfly with enough poison to protect it for its entire lifetime. Apparently most predators learned this lesson countless generations ago and know that this colorfully marked butterfly is a food source to avoid. So repellent are the monarch's distinctive markings that predators will also bypass a non-poisonous butterfly that mimics the monarch wing pattern. Some predators, however, are immune to this toxin and consume monarchs without hesitation.

In recent years Monarch populations have been drastically reduced, particularly since the beginning of this century. One reason for this is the estimated one and a half billion milkweed plants that have been eradicated along the monarch's migration routes. That is an area roughly the size of Texas that has been denuded of milkweeds; the only plants monarch females can use to lay their eggs and raise their larvae.

Pioneer Species
Showy and common milkweed are an early-stage successional species that build habitat for more fragile plants to follow. In this role, they can be found pioneering on disturbed areas resulting from natural events or such human activities as cultivated fields, roadsides, fence rows, irrigation ditch banks, and beside bridge and railroad trestles—another reason, milkweeds might be viewed as a weed rather than a useful plant. A more enlightened way to think of these milkweeds would be as aids to reclaiming degraded lands. Care must be taken, however, to keep it well away from rare, endangered, or small populations of native plants, due to milkweed's ability to quickly colonize and possibly out-compete important species.

Self Defense

Apparently the plant produces its latex sap as a defense against marauding ants. When the ant's feet pierce the brittle surface, they become attached to the white sticky mess and the ant's mobility is quickly curtailed, along with its future.

Native Range and Habit for Showy Milkweed:

Showy milkweed inhabits an extremely large geographical area. In Canada, it ranges from British Columbia to Saskatchewan. In the United States, it is found from Washington to Minnesota in the north and from California to Texas in the south. From British Columbia to central Oregon, it occurs primarily east of the Cascade Mountains, although it can be found in the Willamette Valley. From southern Oregon to southern California, it grows in the coast range and on both the western and eastern slopes of the Sierra-Nevada mountain range. Showy milkweed is also found on the western and eastern slopes of the Rocky Mountains, though it may be less abundant in the extreme southern part of the range. While showy milkweed is a rather common plant, its occurrence throughout its native territory is sporadic.

Showy milkweed grows in the plains and prairies, valleys and bottom lands, and occasionally in mountain parks. It is common in grasslands and also found in open woodlands and forests. Often it grows on the edges or openings within conifers, particularly piñon/junipers and ponderosa pine. In the Pacific Coast Range, it may be found in mixed evergreen forests with both conifers and broadleafs.

Showy milkweed is found along creek banks, in meadows, and among rocks. It might be found near sea level or at altitudes up to 9,000 feet. The most common elevation range in Arizona seems to be between 4,000 and 8,000 feet, however in California, it is rare above 6,000 feet.

Native Range and Habitat for Common Milkweed:

Common milkweed is native from Saskatchewan and North Dakota to Nova Scotia and Maine, down to the southwestern edge of its range in eastern Oklahoma and Texas east to northwest Georgia. In the Southeast especially, common milkweed is primarily a plant of the uplands. In places where common milkweed and showy milkweed overlap the two species readily hybridize.

In the wild, common milkweed is found on forest edges, along riparian corridors, on the banks of lakes and ponds, and in prairies. It is also quite comfortable growing in pastures, dry fields, old orchards, waste places, and roadsides.

CUCURBITA FOETIDISSIMA
(Buffalo Gourd, Fetid Gourd, Chichicoyote, Calabaza, Calabazilla Loca, etc.)
[Cucurbitaceae Family]

Buffalo gourd is a fast growing, non-woody, perennial vine. Each plant usually has six to twenty vines that grow 10 to 25 feet long and tend to hug the ground. The foliage and inner fruit smell like a pile of very sweaty, dirty socks—rank to the point of nausea. That said, this is a handsome plant seen against the barren landscape that it favors. The leaves are robust and triangular, between 4 and 12 inches long, are grayish-green or grayish-blue in color, and borne alternately. The large golden-yellow solitary flowers are beautiful, between 2 to 4 inches in diameter, and look like squash flowers. They bloom May to July. Its spherical fruit is 2 to 4 inches in diameter with light and dark green stripes that mature to a soft gold. There are about 200 to 300 seeds per fruit. These vines have been known in some cases to live over forty years.

Culture:

The buffalo gourd prefers full sun and sandy or gravelly soil. Good drainage is essential. It is extremely drought tolerant due to its huge root. Under arid conditions, the vine can yield up to a ton of seeds per acre. First frosts kill everything

Cucurbita Foetidissima

above ground, leaving only the golden fruit behind. The root is hardy and can withstand temperatures as low as minus 25° F. It is propagated by seeds or cuttings.

While honey bees are likely fond of the protein and carbohydrates that the buffalo gourd flowers offer, the plant relies on wild solitary bees from the genus Peponapis and Xenoglossa for reliable pollination.

Food:
Seeds

The seeds can be used as food in various ways once the bitter and toxic glycosides and saponins have been removed. This can best be accomplished by washing them in mineral-lime water. After shelling them, they can be roasted and eaten or ground into nut butter. Carolyn Niethammer's advice in *American Indian Food and Lore* is to dry the gourds until the green is almost gone. This may take three to five months indoors. Cut the gourds open and separate the seeds from the pulp. Dry the seeds in the sun. Then fry the seeds in a little oil, drain, and salt.

Buffalo gourd seeds are 30 to 35% protein and contain 30 to 40% oil. Due to their high oil content, the seeds are pressed to make an edible oil that can be used for cooking. Puebloan peoples liberated the oil by grinding the seeds and putting the meal in boiling water. After it was cooked, the meal was filtered out and the oil was left floating on the surface where it could be skimmed off and used for cooking. The meal is edible and is reputed to contain five times more protein than sunflower seeds. The meal is often fed to livestock.

Roots
The roots of buffalo gourd can dive 6 to 15 feet deep, have a circumference of 4 feet or more, and weigh as much as 320 pounds. By dry weight the roots are 52 to 55% starch. The starch is extracted by soaking the roots in a dilute salt solution, then processed to remove the bitter toxins. The starch can be used as a sweetener, a stabilizer, or a thickener.

Flowers
The flowers were eaten by the Hopi people, who stuffed them with a cornmeal preparation and then baked them. Other tribes used them for soups and would dry and store them for future use.

Medicine:
The buffalo gourd's nauseating odor results from its cucurbitacin content. This chemical has been shown to inhibit the development of some types of carcinogenic tumors.

Seeds
In his book *The Green Pharmacy*, Dr. James Duke points out that like the pumpkin seed the buffalo gourd seed is a rich source of the amino acids alanine, glycine, and glutamine. In one study, forty-five men took 200 milligrams of each of these three amino acids daily, significantly reducing the symptoms of benign prostatic hypertrophy (BPH).

Roots
Several tribes of the American Southwest used this plant as a medicine. A liquid extract was made by boiling the root then used to treat chest pains, toothaches, earaches, fevers, and constipation. The root extract can speed up childbirth by causing uterine contractions.

Flowers

The seeds and the flowers when mixed with saliva have been used to reduce swelling.

Fruit

The fruit was used to treat rheumatism. It was baked, then cut in half and rubbed, while still hot, on the affected area. The roots may also have been ground and mixed with a vegetable oil then massaged into the affected area.

Warnings and Considerations:

Some of the constituents of buffalo gourd are quite toxic and can prove fatal if taken internally without proper processing. Death is not likely, however, one can become very ill, so it is important to take special care in preparation. Pregnant women should probably avoid using it. Another hazard are the fine, almost invisible hairs on the gourds. If they puncture the skin, they can be painful and are hard to dislodge.

Agricultural Use:

Crop Pests

In his book *Gathering the Desert,* Gary Nabhan tells us that the buffalo gourd has both pest repellant and pest attractant qualities. One Cochiti farmer along the Rio Grande in New Mexico crushes the gourd in water and then wets the foliage of his squash plants with it. This is said to repel squash bugs which buffalo gourds are highly resistant to.

One common application is to drive off bedbugs. The roots are ground and added to water. The water is then sprinkled throughout the bedroom. It is said that the bed bugs immediately leave the premises.

Another method said to keep a home free of insects is to simply cut up a gourd and place a mixture of the gourd and its leaves in each corner of the house.

Buffalo gourd is also reputed to attract Luperini beetles that lay waste to farm crops. In *Gathering the Desert*, Gary Nabhan suggests that buffalo gourd could make an ideal trap plant to lure the pesky beetles away from valuable crops. Apparently these little beasties cannot resist the bitterness of the buffalo gourds.

Other Uses:

Soap

The root and the gourds are rich in saponins which produce a soapy lather for cleaning when rubbed. Clothing washed in it not only comes clean but is also bleached as though it contained commercial bleach. A thorough rinsing is important since bits and pieces stuck to the fabric can irritate the skin. Buffalo gourd saponins are also used as hair shampoo and are reputed to make hair grow. Gourds are cut in half and used to scrub pots, pans, and dishes or to clean up grease spots on wooden floors. Typically the gourds and the root are cut up into little pieces for use as hand and laundry soap as well as shampoo. Large quantities can be stored for future use.

Biofuels

Buffalo gourd's root starch can be used to produce ethanol, and its seed oil can be used as biodiesel. In David Blume's book *Alcohol Can Be a Gas*, he calculates that an acre of buffalo gourd could yield over 110 gallons of biodiesel annually on marginal lands with droughty conditions and poor soils. Blume adds that it is smart to add some of the fatty, acid-rich seeds to the mash because they speed up fermentation.

Fuel for Heating and Cooking

In many arid lands, local native fuels are inadequate to meet the demand for heating, even in rural, low-population areas. Poverty in many of these arid regions is commonplace and for the poor the options are limited. Often all the local trees, if there were any to begin with, were burned for fuel long ago. In many cases wood or coal must be imported, creating a hardship for poor people.

In addition, due to inadequate venting from old leaky stoves and chimneys, pneumonia and bronchitis are uncommonly high in those that rely on wood for heating or cooking, especially among women and children. Years ago, a researcher from Washington University in St. Louis found that buffalo gourd roots burn cleaner and more efficiently than wood or coal. Once sun-dried, they are reputed to burn as hot and longer than many hardwoods. These attributes can significantly reduce respiratory disease rates.

Although considerable effort is needed to liberate the root from the ground, the dried gourds work almost as well for fuel as the roots. They lack the volume of the roots, but they are far easier to harvest and are prolifically produced.

In arid regions, the variety of resources buffalo gourds offer could not only serve the poor, but could actually help relieve a moderate number of refugees from the effects of climate change. Even instances of genocide, like those in the Congo, which are said to be related to the loss of resources caused by a warming planet, might be reduced.

Décor
The gourds are dried and painted as novelties for house décor.

Native Range:
The buffalo gourd grows from Southern California east to central Texas, north to Colorado and Nebraska, and south into Mexico.

EUCOMMIA ULMOIDES
(Hardy Rubber Tree, Du-zhong, Mu-mien, Tu-chung)
[Eucommiaceae Family]

Eucommia ulmoides is the only surviving member of this family and is nearly extinct in the wild. It is a deciduous tree that grows at a moderate rate, reaching 40 to 70 feet tall and is just as wide or wider. The flowers are reddish brown and inconspicuous. The leaves are large and leathery, up to 3 inches long, somewhat resembling elm leaves. Its fruits are encased in a flat-winged nut somewhat like an ash. The bark is grayish-brown.

Millions of years ago Eucommia, like the gingko and the dawn redwood, was native to North America. Also like those two trees it is now extinct, except in Eucommia's case, for tiny populations found in south central China. Propagated from these rare survivors, Eucommias today have been planted throughout much of the world's temperate climate, but it is not commonly planted in the United States because it lacks showy flowers or dramatic fall colors, and its usefulness is mostly unknown. Nonetheless, it has been planted as a shade tree in the United States and does very well.

Culture:

Eucommia prefers full sun and grows at a moderate rate with irrigation, but it does tolerate drought, though the growth rate is reduced. It is very soil adaptable but prefers deep, well drained, sandy or loamy soil, with plenty of humus. It is

Eucommia Ulmoides

pH adaptable and easy to grow as well as transplant. Eucommia is not susceptible to any seriously debilitating pests or diseases. This, however, could change now that its being monocropped, as it has with other hydrocarbon rich plants like milkweed and jojoba. A polyculture system is recommended. It is cold hardy to zone 5 and warmer parts of zone 4. It has survived minus 20° F.

The Chinese grow it on hillsides and other marginal lands unsuitable for most agriculture. The ideal site is a southern slope in the thermal belt. Hillside plantings can benefit from terracing or swaling to increase precipitation capture and to make harvesting easier. Male and female flowers occur on separate trees and are wind pollinated. This suggests males should be planted up wind of females. Leaf production can be more than doubled by hard pruning. Pruning has also been found to increase the nutrient density of the foliage.

Food:

Leaves

In Li Shin-Chen's sixteenth century *Pen Ts'ao* (Materia Medica), he states that consumption of the immature Eucommia leaves was believed to dispel malodorous gas and act as a deterrent to hemorrhoids.

Today in China's northern Hupeh province, the tasty, delicately flavored young leaves are cooked as a vegetable known as mein-ya. Contemporary Chinese still believe that this dish helps expel gas and that the leaves' astringency helps prevent and treat hemorrhoids.

The leaves are also used for a delightful tea called du zhong or tochu that, according to TCM, serves to bolster the immune system, maintain normal blood pressure, and strengthen the legs and lower back.

Bark

The Chinese today use the bark to prepare a reputedly savory tonic cooked like a soup with chicken or pork for a nourishing preventative meal. Some people, however, may find the taste of the bark somewhat less than desirable.

Medicine:

General Supplement

Traditional Chinese medicine has recognized both the preventative and healing powers of Eucommia ulmoides since antiquity. It has been used as "a nutritional supplement" for more than 3,000 years, primarily as a strengthening agent for the liver and kidneys.

Eucommia has numerous antioxidants and many active therapeutic compounds. Pinoresinol diglucoside is the primary antihypertensive agent in Eucommia. It is also rich in flavonoids, including quercetin which induces significant antiviral activity, promotes an anti-inflammatory action, and may help inhibit tumor growth. Additionally, quercetin scavenges free radicals and suppresses oxygen ions that cause tissue damage and foster inflammation. Also present in Eucommia is kaempferol, another antioxidant-rich flavonoid. Other potent antioxidant compounds in Eucommia include chlorogenic acid, geniposidic acid, ferulic acid, caffeic acid, and syringaresinol diglucoside.

According to David Hoffmann, in his book *Medical Herbalism*, in research with mice, ferulic acid has shown the capacity to boost phagocytosis activity which consumes cellular debris in the blood stream or tissues. The caffeic acid content provides analgesic, antibacterial, antifungal, anti-inflammatory, and antiviral effects. Eucommia's lignin glycosides are antibacterial, antifungal, and have demonstrated antitumor and antiviral properties. The lignin furanofuran is also present and works as a antihypertensive.

Other compounds found in Eucommia include aucubin, an iridoide glycoside, and many other aucubin derivatives. This compound has diuretic and laxative qualities. All these constituents are richest in the leaves and bark.

Eucommia is effective as a therapy for moderately elevated blood pressure levels, though less so for high levels. In addition to this cardiovascular effect, it is used to treat cerebrovascular disease.

Eucommia's best medicinal function may be its action as an adaptogen. Adaptogens are herbs that increase the body's ability to resist environmental stress.

Bark

Today Eucommia bark is recognized as an effective treatment for osteoarthritis. This is due to its ability to nourish the connective tissue, cartilage, ligaments, and tendons. For all the reasons Eucommia bark treats osteoarthritis effectively, it is also prescribed for osteoporosis.

Many women entering menopause become victims of osteoporosis, which puts them at a greater risk for fractures or broken bones. While Eucommia is a safe, effective treatment, it is not clear whether it can correct a faulty thyroid gland bent on consuming calcium from the body and the bones, which is often the reason for osteoporosis.

According to TCM and Chinese folk medicine, Eucommia can bestow significant benefits for pregnant women. It can help prevent miscarriage. It can aid in correcting a fetus that has dropped too low as a result of a leaky placenta. It acts as a sedative for a restless, over active fetus. It provides relief from cramps and an aching back, and it is valued for normalizing hormonal imbalances.

The Chinese also use Eucommia to stimulate the

respiratory system, to reduce high levels of cholesterol and triglycerides, to combat dizziness resulting from liver conditions, to strengthen muscles, to ease emotional disorders, to eliminate ring worm, to promote a healthy prostate gland, as a sexual tonic to fight impotence and relieve moist itchy genitals, to treat hemorrhoids, and to increase stamina and speed recovery from fatigue.

Animal research indicates that Eucommia could be a potent anti-obesity herb. Although it is as yet unclear how it works, it seems to burn off fat or its intermediate molecules, and does so remarkably well even at oral doses as low as 3,000 milligrams.

The most effective dosages for the conditions that Eucommia bark treats are yet to be determined, but based on limited evidence, it is usually prescribed at 3 grams taken three times a day or 5 to 16 grams taken on an empty stomach in two doses—early in the morning and at bedtime. Tinctures, decoctions, powder, and pills may be used. It is now known that the bark and leaves are equally potent and are used interchangeably.

A common Chinese system for harvesting the bark consists of planting trees 3 to 6 feet apart and not pruning them at all. The bark is harvested when the trees are ten years old. (Other systems wait fifteen to twenty years before harvesting.) The bark is only harvested from one side of the trunk as girdling kills the tree. For medicinal purposes, the bark is harvested in spring when its active compounds are at their peak. When the bark is harvested, it is stacked and folded so the inner surfaces face each other. The piles are placed in the sun and covered with straw. After about a week, the inner surfaces become nearly black. At this point, the bark is pressed flat and the coarse outer bark is removed. The medicinal potency is increased when the bark is roasted.

Warnings and Considerations:

Eucommia has extremely low toxicity, with no harmful side effects reported.

Agricultural Uses:

Ruminant livestock can use Eucommia foliage for fodder, and when coppiced, it makes for easy animal foraging. Coppicing has been shown to increase the nutritional value of the leaves. Since the leaves are edible and tasty to humans, it would seem this is a good method for producing food for people as well.

Other Uses:

Biofuel, Rubber, Plastic

Hardy rubber tree is a fit name for this truly hardy tree with a significant amount of rubber in its leaves and bark. In recent years it has attracted a great deal of attention as a petroleum substitute. It contains hydrocarbons with an energy value equal to petroleum. Its rubber is a long-chain trans polyisoprene (TPI) found in all parts of the tree except the wood.

Eucommia's rubber has been wrongly confused with gutta-percha, a liquid rather than a solid rubber. Eucommia's rubber is hard and, much like plastic, has thermoplasticity. It is referred to as EU rubber or EU gum. Eucommia is considered high in industrial potential throughout much of Asia, where it is grown commercially. But it can also be grown in much of the continental United States. Considering that most rubber today is artificially made from petroleum—think climate change—and nearly all-natural rubber in the marketplace comes from tropical Southeast Asia, it would make sense to determine if Eucommia could be grown commercially in the United States.

EU rubber is easy to process, is friction resistant, has

superior insulation value, is water resistant, is acid/alkali resistant, and has greater rigidity and a lower melting point than the rubber of the tropical rubber tree Hevea braziliensis. EU rubber's major limitation is productivity. To answer its industrial potential, the TPI yield per tree must increase. Cultivars with higher TPI yields are currently being developed through molecular cloning or metabolic engineering. Improved cultural practices are also being evaluated. (The potential to invade natural systems should be vigorously scrutinized before approval of these kinds of cultivars.)

Despite its low TPI yields, EU rubber has been used for pipe lining, electric and marine cable insulation, heavy wire coating, machine belts, shoe soles, and dental fillings. Its thermo plasticity suggests a host of other biodegradable plastic products.

The bark is harvested for rubber at the end of the growing season when it is most abundant.

Seed Oil

The seeds of Eucommia contain approximately 27% oil. This oil is used both for cooking (it is edible) and for industrial purposes. In addition, it can be used to produce biodiesel fuel.

For commercial scale seed production, mother trees are selected from those that demonstrate prolific seeding. Grafts are made from these trees for bud grafted seedlings.

Wood

Eucommia's wood is used for furniture or firewood.

Native Range and Habit:

Although very rare, wild populations are found in central and south-central China. In the 1970s and 80s deforestation drove them to the very edge of extinction. Reforestation began in the 1980s, but Eucommia is still at great risk in its natural environment.

Eucommia ulmoides is cultivated commercially in Northern China, Taiwan, North Vietnam, South Korea, and Japan for its medicinal properties.

Being cold tolerant, it is planted throughout much of the temperate climate zones of the world as an ornamental or as a shade tree.

FRAXINUS
[Oleaceae Family]

Fraxinus Americana
(White Ash)

White ash is a rapid growing deciduous tree that can reach 70 to 130 feet tall with a canopy width of 50 feet or greater. The symmetrical form is narrow in the forest and broader with an open, rounded crown when given more space. The dark green leaves are 8 to 12 inches long with five to nine leaflets. White ash leafs out in late spring and drops its leaves in early fall. The leaves turn bright to dull yellow or deep purple to bronzy purple at season's end. The fallen leaves break down fast, quickly turning to humus. The purplish flower clusters are inconspicuous, and the male and female flowers are typically on separate trees, although some trees produce perfect (male and female) flowers. The tree flowers before leafing out in late April through May. The tall, straight trunk is commonly 2 to 3 feet in diameter, but in very old age and in ideal locations trunks of 5 to 6 feet in diameter can be found. The bark is an ashy gray, often becoming a dark, grayish-brown. As the tree ages it develops deep, diamond-shaped furrows. Its winged seeds are called samaras.

Culture:
White ash grows best in full sun, but tolerates partial shade. It prefers a rich, deep soil with abundant humus and good drainage. However, it is quite adaptable, doing well on a

Fraxinus Americana

variety of soils, including alkaline soils. Moist soil is a must; neither dry nor soggy will do. In nature white ash is most often found in riparian corridors, frequently on river and stream banks, but it is also found on northern and eastern slopes that are drier than riparian zones but moister than southern and western slopes.

White ash can fall victim to oyster shell scale, ash borers, canker, and flower galls. Young trees are very flammable.

Seeds should be stratified at 40° F for three months and then sown immediately. Named varieties are grafted on seedling rootstock in spring or are budded in summer. White ash transplants easily.

Varieties:

Rosehill is a seedless male variety that is much better adapted to infertile soils and alkalinity than Fraxinus americana and other ash species.

Medicine:

Bark

First Nation peoples found various ways to utilize white ash bark medicinally. A water infusion from the inner bark was used by the Meskwaki to wash sores, cure itching, and kill head lice and similar pests. Other peoples chewed the inner bark or combined it with bear fat, then applied it as a poultice to sores and ulcers. The inner bark was also taken to relieve stomach cramps and to promote sweating to help bring down fevers. A decoction of the roots was made to treat colic. The bitter exterior bark was used as a tonic. Often it was brewed into a tea and consumed after childbirth. First Nation peoples also used white ash as a cure for snake bites. The Delawares allegedly cured a rattlesnake bite by drinking a decoction of the buds or bark.

The pioneers adopted some of the tribal remedies in their own folk medicine. In 1916 the bark was listed in the National Formulary as an official tonic and astringent. The inner bark is used as an emetic, diaphoretic or sudorific, diuretic, and a strong laxative.

Propolis

Honey bees gather a sticky material off the leaf buds of white ash and combine it with wax to make propolis, a product with many medicinal applications for humans. In fact, when all the ailments that propolis has been used to treat are listed, one might believe it to be a panacea.

Early research, primarily from the 1970s, showed that in addition to its antibacterial properties, propolis is antiviral, fungicidal, and an anesthetic. It enhances the function of the immune system and has effectively controlled the potentially deadly staphylococcus aureous. This harmful bacteria is a cause for numerous illnesses, including pneumonia (the sixth leading cause of death in the United States), sepsis (a systemic inflammatory response to infection caused by various pathogenic organisms, killing 200,000 to 400,000 patients in the United States each year), and toxic shock syndrome.

Propolis has had remarkable success in treating influenza, ulcers, various types of acne, dry coughs, and sore throats. It is a source of beta-carotene, B vitamins, vitamins C and E, flavonoids, and various other compounds.

Propolis can be harvested using a propolis trap that replaces a beehive's inner cover. To do this, prop the outer cover up enough to let in a little light. The bees will pack the trap (which is translucent) with propolis in order to shut out the light. Once covered, the trap is put in a bag and frozen. Once frozen, the propolis can be removed by jostling it loose against a hard object.

Seeds

The samaras (winged-seeds) were used as an aphrodisiac, suggesting that the practice of pickling the samara seeds may have had an ulterior motive. (See Fraxinus excelcior, page 84.)

Agricultural Uses:

Bee Forage

The pollen of white ash supplies honey bees with a protein-rich food source, but does not yield surplus honey. Worker bees also harvest a sticky, resinous substance from the leaf buds. The tree coats its leaves with the substance to help ward off desiccation. As said above, the honey bees combine this substance with wax to make propolis. Some white ash trees can contribute to the production of copious amounts of propolis. This product is a valuable resource for the hive. They use it as a sealant for cracks or holes and as an adhesive for honeycombs. Propolis also has significant antibacterial qualities that the bees use as protection against epidemics caused by microbes or viruses entering the hive through foreign agents.

Other Uses:

Wood Products

White ash is valuable as commercial timber. It has a tall, straight trunk, is very strong, has tough elastic properties (thus is easy to bend into desired shapes), is shock-resistant, and produces a heavy wood, weighing 42 pounds per cubic foot when dry. The wood is close-grained with soft, reddish brown heartwood and lighter brown sapwood. Common uses include furniture, cabinetry, interior trim, veneer, sports equipment, agricultural implements, tool handles, musical instruments, ladders, railroad ties, fuel wood, crates, boxes, and basketry.

The wood of American white ash and European common

ash are remarkably similar, and since white ash readily coppices, it could be used to make the canes and walking sticks that common ash is known for. One could also coppice for broom and tool handles and have significant resource productivity on a rather small piece of land. (See common ash wood products, page 85 .)

Windbreak
Resistant to storms and high winds, white ash makes a good component for a multi-row windbreak. A row of tall, dense shrubs or small trees that grow close to the ground is needed on the windward side of the ash trees to block wind from going under the canopy.

Dye
White ash bark yields a very fine, clear yellow or tan colorfast dye. Both green and dried bark work, whether simmered slow or boiled.

Native Range:
White ash is found in Newfoundland and Nova Scotia, west to Ontario and eastern Minnesota, and south from northern Florida to eastern Texas. Pure stands of white ash are uncommon. Other species are usually dominant, but it can be found with some useful companions. Canadian hemlock, eastern white pine, spruce, balsam fir, sarvis berry, oak, shagberry hickory, black cherry, hazelnut, sugar maple, sumacs, liquidambar, and sweet bay magnolia are associates with special virtues of their own.

Fraxinus Exelcior
(Common Ash, European Ash, Yggdrasil)

Common ash is a fast growing deciduous tree, 50 to 90 feet tall on average, sometimes reaching 140 feet in ideal locations. It is 60 to 100 feet in width with a wide-spreading rounded crown. It has purple flowers that bloom before leafing out in the spring. The leaves are a glossy dark green and borne from distinctive black buds with seven to thirteen leaflets per leaf. Common ash leafs out into late spring and drops its leaves as fall begins. Young trees have handsome smooth, light gray bark that fissures deeply as they age.

Culture:

Common ash prefers full sun exposure, with a preference for limestone and calcareous soils, yet it is often found in acidic soils, indicating it is pH adaptable. Regular watering seems to be essential, but soggy soil should be avoided and good drainage is important. Forty-one species of insects are attracted to ash species, fourteen of which are sucking insects. The most common, and one of the most destructive threats to these trees, is oyster shell scale, which can be controlled by dormant oil spray. Borers can also be a serious problem. Common ash is hardy in zone 4.

Varieties:

Hessei is a seedless variety that is more pest resistant than all other ash trees. It is very vigorous and has a very strong, vertical trunk, making it an ideal tree for the timber industry.

Food:

Seeds

The very immature winged seeds called "keys" are pickled. There are a variety of recipes. They should be brought to a boil after removing the little stems and adding salt. Once boiling, allow the brew to simmer for five minutes, drain, and repeat to get rid of the bitterness. Take care not to overcook the keys. You want to make them tender, and overcooking makes them tough and nearly flavorless. In addition to vinegar, a variety of condiments have been added including horseradish, garlic, honey, cider, and peppercorns. Use your imagination. Pickled keys are used with other foods as a condiment.

Leaves

In times of scarcity, common ash leaves have been used to adulterate tea, and the French are said to occasionally ferment them into an alcoholic beverage.

Medicinal:

Common ash contains furanofuran lignans, which include an active ingredient that may reduce hypertension.

Leaves

The leaves are a folk remedy for rheumatic and arthritic conditions, including gout.

Bark

The bark was at one time widely used to treat fever, malaria, and constipation.

Agricultural Uses:

Bee Forage

All ash trees seem to be irresistible to honey bees. The nectar is very rich in sugar, but the pollen content is modest. Rarely are enough trees present in the United States to be a good honey producer.

Fodder

In the past, and to some extent continuing into the present, the foliage has been used as a fodder crop. In some parts of Europe they are coppiced or pollarded as a way to harvest the fodder in volume.

Other Uses:

Wood Products

The wood of common ash is so useful that it is cultivated for timber. It is very strong, hard, and elastic. The light, close-grained wood is shock resistant, which is why it is so popular for the manufacture of sports equipment, particularly baseball bats. Common ash is also used to make furniture, cabinetry, agricultural implements, fuel wood, oars, ladders, pulley blocks, etc. The wood also makes a beautiful and highly valued veneer.

It is an excellent coppice tree. Young saplings of two to three years old are harvested for smooth surfaced canes and walking sticks with a silver-like luster that offer a nice spring with each step. Three to six year-old saplings are harvested for handles for shovels and other tools. In the past, common ash branches have been used to make barrel hoops and crates.

Windbreak

Most ash species appear to be storm and wind resistant.

Fiber

The wood and inner bark of common ash is traditionally used for rope-making. After de-barking, logs are soaked in water for two months, after which they are dried. As the log dries, the annual rings detach from each other in thin segments. Each segment can be split again and again until the ideal size for making cordage is attained.

Native Range and Habitat:

Common ash grows throughout Europe and Asia Minor and is typically found growing with English oak and various shrubs.

Fraxinus Ornus
(Manna Ash, Flowering Ash, Mediterranean Ash)

Manna ash is a rapid-growing deciduous tree 25 to 60 feet tall and 20 to 30 feet wide. It is round headed in form and has showy, richly fragrant flowers in dense fluffy panicles that range in color from white or creamy white to greenish white. It blooms from late spring to early summer and is insect pollinated and self-fertile. Its bright green, glossy leaves have five to nine leaflets producing a dense crown of foliage that provides ample shade. In fall the leaves turn a lovely, soft lavender or yellow before falling. It is an excellent biomass producer.

Culture:

Manna ash prefers sun or part shade and a deep, fertile, loamy soil. Good drainage is also recommended, and it is tolerant of alkalinity. Manna ash is fairly drought resistant and tends to do well with moderate water, but it prefers moist (not saturated) soil. Young trees should be watered liberally during the establishment period. Manna ash's tendency to sucker and re-sprout after cutting makes it a good coppice tree. It is hardy to zone 6 down to about minus 10° F. Manna ash can be propagated through its seeds (requires cold stratification), suckers, or cuttings. They can be grafted in spring or budded in summer. (For information about pests, see the common ash segment on culture, page 83.)

Medicine:
Sap
Manna ash varieties are grown commercially for its mannit exudate that is a gentle but effective laxative, mild enough for infants, children, and pregnant women. It may also be taken as a tonic. The flavor is pleasantly sweet with a slight acidic aftertaste. The exudate can be made into a sugar substitute that is used in preserves or as a sweet syrup. It is often eaten as is and is reputed to be quite nutritious.

The sweet sap is white or pale yellow. In the summer, incisions are made in the bark, and a tube is tapped into the wound to draw out the descending sap. In the fall the solidified exudate is harvested with a thin-bladed knife. When removing the manna from the tube, extra care is given to not cut any of the bark away with the knife. This product is called "large flake manna" and represents the highest quality manna and brings the highest market price. Next, as much exudate as possible is carefully scraped off the bark. This product is known as "small flake manna" and is considered a lower grade. At body temperature the solidified manna softens and becomes pliable as well as water-soluble. It is infused in water before being administered. Dosage ranges from a teaspoon for infants to as much as two ounces for an adult.

Agricultural Uses:
Honeybee Forage
Bees swarm around the manna ash flowers in such great numbers that their buzzing is hard to miss.

Other Uses:
Wood Products
The wood of the manna ash is hard, elastic, shock-resistant, warp-resistant, and seldom bothered by insect activity. The wood is used for veneer, sporting goods, tool handles, and fuel wood. All ash tree species tend to form large burls in their old age that are highly ornamental, each with its own unique patterning. For this reason they are prized by cabinetmakers.

Wind Break
Manna ash is quite resistant to high winds and storm damage, making it a good candidate for multi-row windbreaks. When used in alternate rows with evergreen trees, manna ash makes a good year-round wind break. Without the evergreens, they provide protection only during the growing season.

Native Range:
Manna ash is found in southern Europe and Asia Minor.

GLEDISTSIA TRIACANTHOS
(Honey Locust)
[Leguminosae Family]

Honey locust is a moderate to fast growing tree that is 35 to 75 feet tall and 40 to 50 feet wide. It can grow to a height of 140 feet in the wild, although this is rare. The trunk diameter is 2 to 6 feet. On a good loamy, well-drained soil with irrigation, it can grow 2 feet or more each year. On dry, sterile, gravelly soils expect around 1 foot of growth annually. Honey locust has feathery bright green compound leaves that turn brilliant yellow in the fall. It leafs out late (missing late frosts) and drops its leaves early. The foliated canopy often lets a lot of light through. Understory plants may thrive under the canopy, even growing right up to the trunk. The fragrant, greenish or whitish flower clusters bloom briefly in May or June, usually after the last frost. The flowers on most trees are both imperfect (either male or female) and perfect (having both sexes). For this reason many trees are self-fertile. Pollination is done by insects. Honey locust trees also produce large mahogany pods 1 to 1½ feet long by 1 to 1¼ inches wide. These pods tend to be curved or twisted and contain seeds packed in a sweet pulp. Some trees begin producing pods in four to five years while others may take seven to ten years. The pods are typically borne in clusters of two or three. The tree bears large, thick thorns up to 6 inches long on the branches and trunk. A more wickedly thorny tree is hard to imagine. Some thornless varieties become thorny after forty or fifty years, while some thorny trees will lose their thorns as they mature.

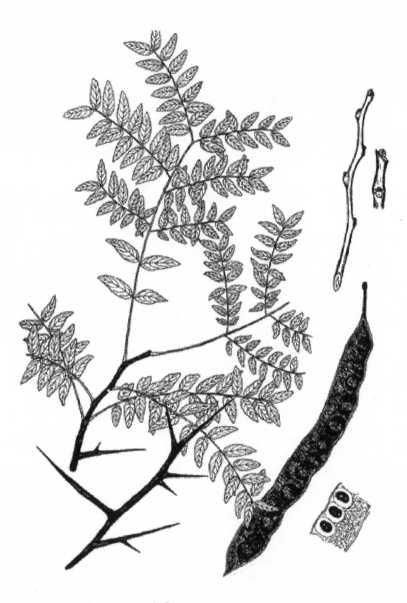

Gledistsia Triacanthos

Culture:

Honey locust trees prefer full sun and a deep, rich, well-drained soil. They thrive in sandy loam and tolerate gravely soils or clays that drain well. In fact, they will grow in almost any soil that has good drainage. Honey locust is moderately tolerant of alkalinity, salinity, and performs well in both acid and alkaline soils. In semi-arid regions, honey locust can survive on just 14 inches of annual precipitation without irrigation. With irrigation it will grow faster. Despite the tree's remarkable drought tolerance, young plantings need copious amounts of water to get established and develop deep tap roots.

Honey locust is relatively resistant to insects and is disease free, although in cities where honey locust may have been overplanted due to its high adaptation to urban conditions, problems of all sorts have occurred. It is tolerant of air pollution, high temperatures, wind, ice storms, and extreme cold.

In addition to the resources honey locust offers, it is widely adapted. It is hardy to zone 4 and is reported to survive temperatures of minus 29° F. It grows well in much of the temperate zone and can also adapt to semi-tropical climates.

Since honey locust almost always has some perfect flowers, male pollinators are not required to produce pods; however, they are important for good seed development. According to an article by agro-forester Andy A. Wilson, one male is sufficient to pollinate ten to thirty females. He recommends thornless males to increase the likelihood of producing thornless offspring. Wilson also suggests that pruning may induce annual bearing in those trees that only bear biannually.

The trees in a honey locust grove are often planted 15 to 20 feet apart. When the grove gets too dense, every other tree can be harvested for timber or fuel wood.

Honey locust is a legume, but it lacks nitrogen fixing nodules. Nitrogen is essential to life and is a constituent of all plant and animal tissues. The mystery of honey locust is how it increases soil fertility and how it can be so rich in protein without sharing the mutualistic relationship with the rhizobium bacteria that makes nitrogen fixation possible.*

For quick germination seeds must be scarified. When soaked for an hour or two in sulfuric acid, the seeds are sufficiently scarified to germinate freely. Sulfuric acid can cause severe burns, so caution is advised. A slower, less reliable method of germination involves soaking the seeds in hot water.

Cuttings of named varieties can be grafted onto rootstocks, or they can be budded on rootstocks that are still in the seedling stage. Root suckers can also be severed from the mother tree and planted. In many cases the thornless trees can be the most desirable. Naturally occurring thornless trees will breed true, ensuring thornless offspring. However, some cloned thornless varieties have been developed by taking cuttings from the thornless upper branches of desirable thorny cultivars. These trees will not breed thornless offspring.

Varieties:

Thornless honey locust trees were originally developed for landscaping, but many of these thornless varieties are intentionally poor pod producers, while others are highly productive.

Bill Mollison once told me that he thinks they may be noduated at a deeper level than has been previously explored.

Ashworth: This variety has a distinctive flavor with melon-like sweetness. The abundance of male flowers makes it a good pollinator. It also has a very high cold tolerance and is thornless.

Calhoun: Calhoun is another very sweet variety with up to 39% sugar content. Yields of 30 to 35 pounds of pods by the age of five years are not uncommon.

Halka: Halka is fast growing. This variety has a very sturdy structure, and it is a good producer.

Inermis: This is the original mother of thornless honey locust cultivars. It has yielded up to 400 bushels of pods per acre (about 3,200 gallons). It is reported to bear annually, with sugar and protein contents of 29% and 20%. These sugar and protein levels can be maintained for up to three years in storage.

Millwood: Millwood has large, super sweet pods with 37% sugar content. It grows rapidly, 2 to 3 feet annually, even on poor soil. A single tree yields 60 to 65 pounds of pods after five years of growth. It is remarkably hardy to as low as minus 30° F and is thornless.

Nana: Nana offers high pod yields.

Schofer: This very cold resistant variety produces pods of the same size and sweetness as the millwood variety.

Dozens of new varieties are still being developed.

Food:
Honey locust could become an important source of food, but it has some compounds that are potentially toxic (see the "Warnings and Considerations" section).

Pods

The pods are a rich source of sugar, starch, protein, calcium, and iron. A typical pod may contain 14 grams of protein, 2 grams of fat, 60 grams of carbohydrates, and 18 grams of fiber.

Tender, immature seed pods are eaten cooked. Seed pods may be ground for meal and eaten as a gruel, and fresh pods can be ground and fermented in water to produce a strong bitter-sweet beer. The moderately sweet immature seeds taste similar to raw peas and are eaten raw or cooked. The toasted seeds are used as a coffee substitute. The seeds can also be soaked in water and sprouted for a side dish.

Many people find the sweet succulent pulp delicious, although the pulp in older pods becomes bitter. It can be used to make a really tasty relish.

The pulp has been converted into a sugar substitute, but it is commercially impractical because of the pods' low pulp content. The sugary pulp can be fermented to brew a unique alcoholic beverage that was historically called metheglin.

Medicine:

Bark

The Cherokee and the Delaware peoples used the inner bark, combined with the bark of the sycamore, to brew a tea that they gargled for sore throats and hoarseness. An antiseptic tea prepared from the pods was prescribed for inflammation of the mucus membranes, indigestion, and whooping cough. The Fox used it to treat measles, small pox, colds, and fever, while the Creak made measles and small pox remedies by boiling the branches, thorns and all. The Rappahannock used both the bark and roots for cold and cough relief. The Delaware used the bark as an antidote for coughs and as a blood purifier.

Pods, Pulp, and Leaves

The Cherokee used the pods to treat measles and dysentery. The pod pulp contains the alkaloid stenocarpine which has been used as a local anesthetic that also contains antiseptic compounds. Russian scientists are looking into honey locust's leaves as a treatment for certain types of cancer (Duke and Foster, *Medicinal Plants and Herbs—Eastern/Central*).

Warnings and Considerations:

All parts of the tree contain potentially toxic compounds. Therefore, it may be wise to ingest them in moderation, at least until its toxicity is better understood. It is thought to contain the alkaloid atrophine, which as a salt is called atrophene sulfate. Medicinally, it has been used to treat potentially life-threatening bradycardias and heart blocks. On the other hand, atrophine sulfate poisoning can cause a variety of symptoms, including abnormally rapid heartbeat, hallucinations, delirium, coma, and other neurological side-effects. With breeding, the toxic principles could be reduced or eliminated, as has happened with many common edible plants that originally contained toxins. It is possible that cooking may render the toxins harmless, but this needs further investigation.

Agricultural Uses:

Erosion Control and Living Barriers

Due to honey locust's deep and extensive root system it is valuable for erosion control, as it has clearly demonstrated in reclamation plantings (see page 99).

When planted densely, it can also serve as a living fence to keep wildlife predators out and livestock in. Its thorns acts as a formidable deterrent.

Bee Forage

Honey locust is a rich source of nectar for honey bees and other beneficial insects. The short blooming season, however, is not long enough to support yield surpluses or honey, but it is long enough to support beneficial insects so they stick around to provide pest control.

Livestock

Animals love the pods and the foliage of the honey locust tree. Both are rich sources of protein. The foliage alone contains up to 30% protein content. Once the pods reach maturity, cattle and hogs will patiently wait beneath the canopy for them to fall. When pods are ground up, they are easier for animals to digest and their nutrition is enhanced. On a diet of honey locust pods cows will produce milk that is richer in butterfat and hogs will fatten up nicely. On a pound-for-pound basis, the ground pods are equal in feed value to oats.

A fodder can be made with a hammer mill. To grind the sweet, sticky pods in a hammer mill, a dry meal such as bran is mixed in to improve mixing and handling. Sheep are possibly even better adapted to pasturing among the honey locust than cattle. Honey locust's open canopy allows pasture grass to grow right up to the tree's trunk. However, in maturity, the canopy can become denser and the branches can droop down close to the ground, barring easy access beneath the canopy.

Companion Plant

The genus Lespedeza can make an excellent companion to honey locust. Lespedeza bicolor (bush clover—see page 100) enhances honey locust's erosion prevention capacity. Honey bees work bush clover flowers for pollen, an added benefit to the bees that avidly gather the abundant nectar of the honey locust. The nectar supplies the bees with carbohydrates while the pollen supplies protein, providing a complete diet.

Livestock can also benefit from bush clover's nutritious high-protein foliage. Unlike honey locust, bush clover is very well nodulated with nitrogen-fixing bacteria. With the exception of the coldest parts of its range, bush clover also produces a high BTU fuel wood. These two plants can have a highly symbiotic relationship from which both may benefit.

Other Uses:
Ethanol
Honey locust may prove to be among the very top ethanol producers worldwide. With the sugar content of its pods reaching as high as 39%, fermentation is easy. The millwood variety contains 36% sugar. Five-year-old millwood varieties have average yields as high as 58 pounds of pods per tree. At forty-eight trees per acre, these trees can produce almost 2,800 pounds of pods. On drier, harsher sites twenty trees per acre may be more practical. Some mature trees can yield 400 pounds of pods per tree, and yields of 1,000 pounds have been reported.

Both the genetic makeup of individual trees or varieties and site conditions affect yields. Improvements in yields and sugar content will result from breeding efforts and identifying trees that are well adapted to local conditions. From these mother trees many varieties can be bred for quality, productivity, and local conditions.

Wood Products
Honey locust wood is strong and hard, weighing 42 pounds per cubic foot. It has a vivid yellowish brown color with a dark brown mottling. The sapwood is a lighter color. The

wood is used in general construction, cabinetry, finish work, veneer, furniture, and fencing. The wood is very durable in contact with the soil and makes long-lasting fence posts. It is also a popular long-burning, high BTU fuel wood. The wood is particularly valued for grilling because of the delicious sweet-smoky flavor it bestows to foods.

Honey locust trees tend to coppice vigorously, so that with a bit of training, they will grow more rapidly into a tree than from ordinary planting. This is due to the fact that it does not have to develop a new root system—because it already has one.

Windbreaks

The honey locust is exceptionally tolerant of high winds due in large part to its taproot and its deep rooting nature. It is used in multi-row windbreaks but must have a windward row of tall shrubs densely foliated to the ground to avoid problem wind from ducking under the honey locust's canopy. Such problem winds increase in velocity due to the venturi effect and can devastate crops. In extreme winds honey locusts can lose some branches.

Shelterbelts

During the dustbowl, President Franklin D. Roosevelt initiated a shelterbelt program designed to control erosion, reduce wind velocity, and increase soil moisture by catching and shading snow in order to reduce evaporation and slow snowmelt. (This can double and triple the amount of precipitation captured by the soil.) About 200 million trees were planted over a massive area, and these shelterbelts were highly effective at bringing the dust bowl under control. Many different tree species were used, including honey locust which of all the trees planted had the highest survival rate at 79%.

Ecological Functions:

Native honey locust has a valuable reclamation component. It has helped to restore and protect semi-arid lands worldwide. The extensive root system and remarkable wind resistance of these trees make them an ideal choice for reclamation and erosion control. Some of the many animals that utilize honey locust for food and shelter are deer, rabbits, squirrels, and quail.

Native Range:

Honey locust can be found from New York to Florida and west to South Dakota and Texas. They are also found in Southern Canada.

LESPEDEZA BICOLOR
(Bush Clover, Yama-Hagi)
[Leguminosae Family]

Bush clover is a fast-growing deciduous shrub 4 to 12 feet tall, erect, and round-topped. It bears an abundance of rosy purple, pea-like flowers in erect panicles. It blooms from August to September over a fairly long period. The tri-foliate, clover-like leaves with blunt oval leaflets are connected to slender, wiry stems. It is commonly found living in small thickets.

Culture:
Bush clover is easy to grow, enjoying full sun with a preference for coarse, sandy soil. Well-drained soil is essential. It is quite tolerant of poor soil and is pH adaptable. Bush clover does very well in average garden conditions and has no serious pest or disease problems. It is hardy to zone 4, dying to the ground in the coldest climates but re-emerging in the spring. Bush clover is often recommended for its usefulness in upland tropics and subtropics, although it is native to the temperate zone. It can be propagated by cuttings. Bush clover is noninvasive.

Related Species:
Lespedeza thunbergii is another rapid-growing shrub very similar to Lespedeza bicolor, but more ornamental and lacking its high fuel wood properties. It is hardy to zone 5 and also pH adaptable.

Lespedeza Bicolor

Food:

Edible plant expert Stephen Facciola tells us in his book *Cornucopia II – A Sourcebook of Edible Plants* that the tender new growth, both leaves and stems, can be boiled or fried or can be eaten as greens. The foliage is also used for tea. Sometimes the seeds are boiled for a pulse that is added to rice dishes.

Agricultural Uses:

Forage and Fodder

The high protein content of bush clover foliage serves as a nutritious feedstock. It can be cut and chopped and then fed to livestock much like alfalfa bales, or the animals can simply be allowed to browse it.

Bee Forage

Bush clover is very attractive to honey bees. In areas where bush cover grows abundantly over sufficient acreage, honey bees can produce surplus honey as seen in South Korea and Japan. Honeybees work this plant vigorously, particularly for its pollen, which provides them with protein.

Nurse Crop

The roots of bush clover are well-nodulated, making it a valuable source of nitrogen. In reforestation projects it has been planted with pine seedlings to increase soil fertility. It can also be planted in orchards as a nurse crop for fruits and nuts. It makes an excellent companion to honey locust in shelter belts or on biofuel plantations.

Erosion Control

The bush clover roots can effectively bind soil, making it resistant to erosion.

Other Uses:

Fuel

The hard, dense wood of bush clover is popular as a fuelwood in many parts of the world. It is easy to harvest and burns hot.

Native Range:

Bush clover is indigenous to northeast Asia, northern China, Manchuria, and Japan.

MYRTUS COMMUNIS
(Myrtle, Sweet Myrtle)
[Myrtaceae Family]

Myrtle is an evergreen shrub that grows at a moderate rate, reaching 5 to 6 feet in height and 4 to 5 feet in width. In great old age it may eventually reach a height of 9 to 15 feet and a width of 20 feet, creating a dense, neatly rounded form. The foliage is a shiny, bright green and is very aromatic. Small, fragrant white to pinkish flowers are borne profusely, sometimes as early as June or as late as October, but mostly from July to September. In southern Europe they tend to bloom from May to July. The flowers develop into a small, very dark blue to black fruit up to ½ inch in diameter.

Culture:
Sun to part shade is ideal for the myrtle. Although myrtle is at its best in rich loam, it is quite adaptable to almost any soil with good drainage, including alkaline soils. In southern Europe it often naturalizes on stony ground. Myrtle prefers deep watering that is spaced widely over time. Shallow watering or poorly drained soil results in chlorosis.

Myrtle can, however, be quite water efficient and even on semi-arid sites will tolerate some drought. Myrtle is easy to grow, low-maintenance, and self-fertile. It is tolerant of high heat and is highly resistant to Texas root rot. Shelter it from cold, drying winds. It is hardy to zone 9.

Myrtle can be clipped and pruned for size and shape. It is not only very tolerant of pruning, but also a good biomass producer. Thus, harvesting a quantity of foliage is harmless in

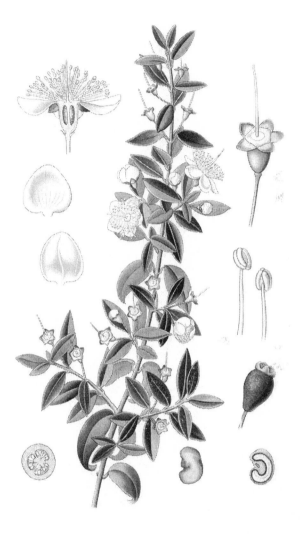

Myrtus Communis

moderation.

Seeds are sown in spring when temperatures reach 60° to 65° F. They should be kept moist and shaded until germination and then transplanted into pots. Clones may be propagated by striking cuttings. Layer in early summer and once rooted, pot it or plant it.

Varieties:
Several varieties have been developed in Europe, mainly for improved fruit and higher yields. Some varieties are good biomass producers. Other varieties are favored for essential oil distillation.

Tarentine is a compact, narrow leaf form of myrtle. It is much hardier than the species by about 5° F or more, and is borderline in zone 7. It is also more wind resistant and makes a good choice for windbreaks and suntraps.

Food:
All the above ground parts of the myrtle are highly aromatic. The dried fruits, leaves, and flower buds are used to flavor poultry and other meats, jams, sauces, and liqueurs. A fragrant essential oil is steam distilled from the fruit, foliage, new shoots, twigs, and bark and can be used to flavor various dishes. The fruit can be used as a condiment or to make an acidic fermented beverage. In Italy the flower buds are consumed as a side dish.

Medicine:
Leaves
Myrtle's distilled leaves are used to treat bronchial congestion (as an antiseptic expectorant), paranasal sinus inflammation, (as an antibacterial), and dry coughs. Distilled leaves are also used as an antiseptic and an antibiotic to remedy urinary tract

infections, vaginal discharges, gum infections, and gingivitis. Due to a very potent antiseptic compound found in the leaves called "myrtol," both internal and external application work as treatments for rheumatism, dysentery, diarrhea, hemorrhoids, gas relief (carminative), and acne (as an astringent).

Agricultural Uses:
Sun Traps

Myrtle is a good choice as a component for a sun trap. A sun trap is a planting shaped much like a horseshoe. Evergreen shrubs, or trees if space allows, are planted on the north side of the garden and wrapped around the garden on the east and west sides in the shape of a horseshoe. On the east side, and more importantly on the west side, lower-growing shrubs are used to allow more sun into the garden. The south side is left open. A sun trap is used to create a microclimate around a garden to either extend the growing season or to help protect a winter garden. Evergreens, unlike woody deciduous plants, do not go dormant over the winter. Hardy evergreens therefore have to heat themselves internally using solar energy to survive the winter. Some of the heat they produce escapes and warms the area around it, generating a warmer microclimate in its vicinity. Depending on the direction of problem winds, a hedge of evergreens like myrtle can protect the garden from wind damage and reduce wind chill. Suntraps, however, can be counterproductive if crops are being grown on north slopes where they can trap cold air drainage in the garden.

Bee Forage

The myrtle flowers are utilized by bees for the pollen. Thus they can be an agent of pollination as well as contributing to honey production.

Other Uses:

Cosmetics, Soaps, Skin Care, etc.
The essential oil of the myrtle, considered the "Eau D'Anges" of scent making, is used for fine perfumes and toilet water. The essential oil is also an ingredient in skin care products, cosmetics, and soaps. In the Mediterranean region, the flowers are still considered the traditional bride's flower, and bouquets are made for weddings. Indeed, myrtle has symbolized love since antiquity, no doubt due to its deeply sensuous ambrosia. The dried flowers are used in potpourri, and the dried leaves are used in herb pillows.

Leather Tanning
The bark and roots are a source of tannins and are used in Russia and Turkey to tan the finest leathers.

Furniture Polish
Myrtle was once used to scour furniture. It also makes a good addition to furniture polish, leaving an antiseptic surface and a delightful fragrance on furniture.

Wood
The wood is hard, very close-grained, and pliant. It is used for furniture as well as products that must be bent or absorb shock.

Native Range and Habitat:

Myrtle's origin is unknown, but thought to be western Asia where it grows wild. It has been naturalized throughout southern Europe, is grown commercially in northern Africa, and is also grown in Corsica, Crete, Sardinia, Russia, Turkey, and the Mediterranean mainland as well as in the United States.

Myrtle is commonly found growing in chaparral among other evergreen shrubs, in open pine groves, or often on stony ground in west Asia where it has naturalized.

ORIGANUM VULGARE
(Oregano)
[Lamiaceae Family]

Oregano is a perennial herb 1 to 3 feet tall, spreading by underground runners 1½ to 3 feet across. It produces white, pink or rosy purple flower clusters from about late June to September. For culinary or medical use, the best varieties are those that are rich in essential oil with a strong, rich flavor. Many varieties have little to no flavor. This is generally true of the species as a whole. (See list of varieties for culinary and medicinal varieties, page 112.)

Culture:
Oregano is easy to grow. It likes full sun, although in regions where summers are very hot, some afternoon shade is recommended. Good air circulation is also an important consideration. It prefers moderately rich, gravelly loam, though it does well in a variety of soils as long as they drain well. It requires little to moderate irrigation. Overwatering can cause fatal root rot. Plenty of compost or organic matter in the soil is desirable and that is about all the fertilization it needs. It is also fairly resistant to pests and disease. Being short-lived, replanting after three years may be required. It is easily grown from seed. For best flavor it should be harvested before budding. Repeated harvesting thereafter keeps it succulent. Flavor-wise, dried oregano is superior to fresh oregano. Hardiness varies a lot with each variety. The most cold tolerant survives zone 4, but many varieties are not that hardy.

Origanum Vulgare

Varieties and Subspecies:

It is wise to avoid purchasing plants vaguely labeled "Origanum vulgare" because they are often low in the essential oil that gives the herb its flavor and medicinal properties. Instead, pick varieties and subspecies known for their rich, spicy flavor, and aroma. Those listed below are among the very best.

Origanum Vulgare Compactum Nanum (Dwarf Oregano): This mat-forming oregano is only 2 to 4 inches tall and spreads to about 2 feet. It produces short purple flower spikes in the summer and has a strong, spicy flavor.

Origanum Vulgare var. Dark (Dark Oregano): Dark Oregano grows uncommonly erect, up to 2 feet tall with whitish-pink flowers. It is reputed to have an excellent culinary flavor.

Origanum Vulgare Subsp. Gracile, Syn. O. Tyttanthum (Khirgizstan or Turkestan Oregano): This is an exceptional ornamental about 1½ feet tall with glossy green leaves and pink flowers. It has a strong, spicy natural oregano taste.

Origanum Vulgare Subsp. Hirtum var. Hot and Spicy (Hot and Spicy Oregano): Hot and Spicy Oregano is a large, vigorously growing oregano, reaching 3 feet in height. As its name implies, it is distinctly hot and spicy. It is claimed to be hardy to zone 5. Others say it is only hardy to zone 6. It has been labeled a "hybrid" by some, a subspecies by others.

Origanum Vulgare Subsp. Hirtum var. Kaliteri (Kaliteri Oregano): This variety is 1½ feet tall with silver-green foliage. It is an important commercial strain in Greece, selected for its extra high essential oil content. It has a delightfully strong, spicy flavor without any bitterness. It is not as hardy as other oreganos, rated for zone 7 to 11.

Origanum Vulgare var. OV12 (Profusion Oregano): This variety is a registered trademark of Richters, all rights reserved. Richters says it is an intensely aromatic and flavorful strain. A colony of OV12 survived over twenty-five years of total neglect while competing with native grasses and herbs, suggesting that it is a very self-sufficient plant. Richters rates it as hardy from zone 5 to 9.

Origanum Vulgare Subsp. Hirtum (Greek Oregano): Greek Oregano is another vigorously growing subspecies that can get as tall as 1½ to 3 feet and as wide as 1½ to 2 feet. It has white flowers and broad, bright green foliage. The strong, spicy aromatic flavor of Greek oregano is considered to be the "true" oregano taste. Although excellent for culinary purposes, its cultivar Kaliteri is said to be even better, but less hardy than its Greek counterpart. Greek oregano is hardy from zone 5 to 11.

Origanum Vulgare var. Seedless (Seedless Oregano): This is possibly a naturally occurring hybrid between Greek Oregano and marjoram. The flavor is very similar to the former, except sweeter and less bitter. It is less cold resilient than Greek oregano, being hardy to just 10° F.

Origanum Vulgare Subsp. Hirtum var. Italian (Italian Oregano): This too is presumed by some to be a hybrid between Greek Oregano and marjoram, sometimes called "Sicilian Oregano." It grows 1 to 2 feet tall and has white flowers with gray-green or pale-green foliage. It has a strong, spicy, and sweet flavor without any bitterness. It is an excellent culinary herb, hardy to zone 7.

Origanum Vulgare var. White (White Oregano or Viride): This variety grows from 1 to 2 feet tall and has white flowers. It has a strong, rich flavor and is another excellent culinary herb.

Food:

Oregano can be prepared fresh or dried, but has the best flavor when dried. Oregano's foliage is an immensely popular culinary herb for soups, stews, casseroles, stuffing, baked goods, relishes, sauces, pasta, olives, vinegar, meats, and dairy products. It is often used to spice up salads, sliced tomatoes, tomato sauces, salsa, tomato pastes, and salad oils. Oregano is used liberally in Mexican, Italian, Greek, and Turkish cuisine. In India the new shoots and foliage are cooked and served as a vegetarian dish.

Oregano's essential oil is used in a variety of ways in the food industry. The ketones and carvones it contains add flavor to sweet treats and liqueurs. The foliage also makes a stimulating tea.

Medicine:

Functions

As popular as oregano is for culinary uses, it is not well known as a medicinal herb, yet it is an extremely rich source of antioxidants, beneficial phytochemicals, and other health-promoting substances.

Oregano has some two-dozen analgesic compounds with several anesthetic properties. It also contains over two dozen anti-inflammatory compounds, at least ten compounds that are antihistaminic, and another ten that are antiallergenic. It is also a rich source of antiherpetic and antiviral compounds.

Eight of oregano's phytochemicals act as natural COX-2 inhibitors (both anti-inflammatory and analgesic). A list of these phytochemicals and their properties follows:

Apigenin: An antibacterial and diuretic that also promotes hypotensive actions.

Caffeic Acid: An antibacterial, antidepressant, antifungal, antihepatoxic, antioxidant, antiulcerogenic, and antiviral.

Eugenol: An analgesic (often used in dentistry) and a carminative.

Kaempferol: A common antioxidant flavonoid.

Quercetin: A common flavonoid that is anticarcinogenic, antioxidant, antipyretic, antitumor, and antiviral.

Rosmarinic: An antibacterial and an antiviral.

Oleanolic and Ursolic Acids: These finish out the eight natural COX-2 inhibitors found in oregano.

Ketones are another group of phytochemicals, four of which are found in oregano:

Camphor: A mild anesthetic with antipuritic and carminative properties (often used to relieve joint pain and swelling in those suffering from arthritis). It also promotes localized blood flow.

Carvacrol: A potent antiseptic and antimicrobial with anthelmintic, antifungal, and carminative properties.

Carvone: An antiseptic and carminative.

Thymol: An antifungal, antiseptic, antimicrobial, and carminative (used to eliminate molds).

Oregano also contains natural monoamine inhibitors that are used in the treatment of depression and Parkinson's disease.

Antioxidant

Oregano's antioxidant content and healthful effects are outstanding. A report published in the *Journal of Agriculture and Food Chemistry* that compared the antioxidant content of twenty-seven herbs, both medicinal and culinary, found that oregano was well ahead of all the others. In another study, oregano and almost 100 other members of the mint family [Lamiaceae] were tested, and oregano was found to be by far the highest in antioxidants. A single tablespoon of fresh oregano contains at least forty times the antioxidant potency of an apple.

A USDA analysis of antioxidant sources found that oregano's antioxidant content was even greater than fruits and vegetables that were known to be exceptionally high in antioxidants. For example, oregano has a four times greater antioxidant content than blueberries.

Antioxidants prevent or inhibit the oxidation of cells by free oxygen radicals. These destructive radicals are highly reactive chemicals that can speed up the aging process and cause cell disease. They can even cause some forms of cancer.

According to research, the rosmaric acid found in oregano is largely responsible for much of its antioxidant properties and also contributes to its antibacterial, anti-inflammatory, and antiviral qualities as well. Consuming high concentrations of antioxidants can help sustain the immune system of those afflicted with HIV. Oregano could even prove helpful in preventing the onset of AIDS.

Antimicrobial

French physicians researching the antimicrobial effects of herbs found that oregano stood well ahead of the pack. Out of several hundred types of microbial infections, oregano extracts either successfully controlled or helped to improve 85% of them. It should come as no surprise that oregano contains nineteen antibacterial chemicals that control numerous bacterial pathogens.

In Mexico, researchers found that oregano's antimicrobial action proved more effective in treating Giardia lamblia infections than the popular prescription drug tinidazole. Oregano's antimicrobial activity can help control impetigo caused by Pseudomonas aeruginosa and Staphylococcus aureus.

Antiseptic

Oregano is a rich source of antiseptic compounds that suppress infection-causing bacteria. Carvacrol, found in the essential oil produced from oregano flowers, is highly antiseptic and appears to contribute significantly to oregano's ability to fight infections. Commercial mouthwashes often include oregano's essential oil for its potent antiseptic action. The carvacrol content may also act as a vermifuge that kills or ejects intestinal worms.

Chewing oregano leaves for temporary toothache relief is a time-honored folk medicine practice. Its effectiveness has been credited to its antiseptic compounds, but perhaps its analgesic, anti-inflammatory, and antioxidant properties provide further support.

Antifungal

The essential oil of oregano is also a very potent antifungal.

Various studies show that oregano effectively controls numerous fungal and yeast pathogens, including many that are food-borne.

Analgesic
Oregano is analgesic (relieves pain). It is typically prescribed to provide relief for sore muscles, stiff necks, nervous headaches, menstruation pain, and even chronic pain as in osteoarthritis.

Inflammation can aggravate existing pain or produce pain itself. Oregano's anti-inflammatory properties may work in tandem with its analgesic effects. Inflammation is often a byproduct of free radical damage, suggesting that oregano's antioxidants also play a role in pain relief, at least where cell damage by oxygen radicals is involved.

The leaves of oregano are used in topical preparations that provide relief from skin irritations such as rashes, rheumatism, and swellings. It also serves to relieve itching.

Antiviral
Due to oregano's antiviral properties, herbalists may recommend it to treat the common cold, coughs, bronchitis, measles, mumps, tonsillitis, and other viral conditions.

Miscellaneous
Oregano has seven compounds that reduce high blood pressure, four antiasthmatic substances, and six expectorant compounds.

Warnings and Considerations:
Do not ingest oregano during pregnancy or lactation. It is also important not to use the essential oil externally or internally in its undiluted form because it can cause severe burns,

especially to sensitive parts of the body. To dilute, add just a few drops of the essential oil to a plant-based oil like almond oil or coconut oil. Do not confuse essential oils (many are toxic in their pure forms) with regular plant oils pressed from seed and commonly used in cooking; they are very different. The pure essential oil of oregano, in particular, is an irritant to the mucus membranes. Some people can develop systemic allergies to oregano.

Agricultural Uses:
Fodder and Climate Change
The release of methane is a byproduct of digestion in ruminants like cattle, goats, sheep, buffalo, and bison. Methane is twenty-three times more potent as a greenhouse gas than carbon dioxide.

In recent years a dairy scientist at Penn State, Alexander Hristv, has created a feed supplement made up primarily of oregano's volatile oils, including carvacrol, geraniol, and thymol, which reduces methane emission from dairy cattle by 40% in laboratory experiments.

Reducing methane production in these animals also gave them more energy and improved their milk production levels. Each dairy cow yielded almost 3 pounds more milk daily when given oregano-infused feed.

Beneficial Insects and Pollinating Bees
Oregano flowers provide pollen and nectar (protein and carbohydrates) to predatory and parasitic insects that control pest insect populations. The flowers also supply food for bumblebees and other wild bees critical to the pollination of food crops and native plants.

Insecticide

While oregano flowers host beneficial insects, they also have insecticidal properties and have been used successfully to control fruit flies. Some gardeners plant oregano here and there throughout their garden as a general pest repellant. Empirical evidence suggests that pests may avoid it because it happens to be habitat for predators. On the other hand, it is thought that many pests seek out their targets by smell and that highly aromatic herbs like oregano interfere with their ability to detect them.

There has been no mention of oregano's essential oil being used as a fungicide against plant diseases, but it can control a large number of fungal diseases effectively in mammals.

Other Uses:

Massage Oil and Other Products

While oregano's essential oil can cause severe burns to the skin, when properly diluted with regular vegetable oil (for massage, almond oil is one of the best), it has the ability to penetrate the muscles deeply and spread a soothing analgesic feeling to aching muscles. Massage oil containing oregano's essential oil helps the body relax and let go of stress. In addition, oregano's essential oil is used to make salves, liniments, and various herbal bath products.

Dye

For centuries oregano flowers formed the basis for a reddish-purple dye used to color wool products.

Food Preservative

Being strongly antibacterial and antifungal makes oregano's essential oil a very effective food preservative. It is used to protect grains from Aspergillus flavus and Aspergillus parasiticus. These two fungi grow on stored grains, producing aflatoxins that can cause liver damage. Oregano's essential oil has proven effective in obstructing these fungi's production of aflatoxins as well as suppressing their mycelial growth.

Oregano's flowering tops have been used in the brewing of beer and ale as both a preservative and as a flavoring.

Perfumes, Soaps, and Cosmetics

Oregano's essential oil is used in the production of commercial perfumes, soaps, toothpaste, mouthwashes, air fresheners, potpourris, herb pillows, and moth repellents. The camphor and thymol found in oregano's essential oil are natural preservatives and are used to preserve cosmetics and botanical and biological specimens. The essential oil's antiseptic quality makes a valuable addition to these products.

Native Range and Habitat:

Oregano is native throughout the European mainland, but not the islands, and throughout Asia. In its native range, it favors dry rocky sites on coarse grasslands, hillsides, and open woods. It has been naturalized in some areas of the eastern United States.

PHRAGMITES COMMUNIS
[Syn. P. Australis, P.C. Subsp. Australis, P.A. Subps. Australis, P.A. Subsp. Berlandieri, P.C. Subsp. Berlandieri]
(Common Reed, Giant Reed, Carrizo, Lugen, Nal Danube Grass, etc.)
[Gramineae Family]

Common reed is one of the most ancient species of grass on Earth. It is an extremely rapid-growing, woody, perennial grass 5 to 16 feet tall, occasionally taller, and spreading 17 to 34 feet wide. The common reed has thick, hollow stems (culms—like bamboo) which can reach nearly 1 inch in diameter and possesses stout rhizomes and stolons that can grow into large, dense stands. Flowers bloom in late July or early August, forming large terminal panicles, covered with silky hairs, usually reddish at first then maturing to a soft, purplish-silver haze. These flower plumes tend to be 15 to 20 inches long and can remain in good condition all winter.

Culture:
Common reed prefers full sun and well-drained, infertile soil. It is tolerant of alkalinity, salinity, and limestone soils. It is hydrophilic, requiring moist soil year-round, and often grows in water up to 6 feet deep. It grows from the tropics to the temperate zone and at sea level to 8,900 feet in the Colorado Rockies and 12,500 feet in the Andes. Cold tolerance

Phragmites Communis

depends on the native habitat of the plant material. Those from the coldest zones are hardy to at least minus 20° F and probably much colder. Propagation is by rhizome division.

Food:

Since antiquity common reed has provided an abundance of food for cultures worldwide. Just about every part of this plant has been used for food. The rootstock, stolons, sap, tender young shoots, stems, partially unfurled leaves, and seeds can be eaten. In some parts of the world it is still part of the everyday diet.

Rootstock and Stolons

The rootstocks and stolons are gathered year-round, providing the ground is not frozen. After being washed and peeled, they are consumed raw, boiled, or roasted. A fair quantity of starch can be liberated from the common reed by first crushing several of them in water and letting them soak. The water is then poured through a screen or filter into another container to remove the courser parts. The starch sinks to the bottom. Once it has settled, the clear water is carefully drained off, and the starch is eaten as is or dried for use as baking flour.

Shoots

The tender new shoots that emerge from the base of old stems in the spring can be cut and eaten fresh, steamed, or boiled like a vegetable. It has a rather distinct, sweet flavor. Young shoots are also pickled. The tastiest pickles come from young shoots that have not yet broken through the water surface and emerged into the sunlight.

Culms

The culms (the stems of the grass family) are harvested in early summer before flowering. They are rich in sugars during this period. Once cut, they are sun-dried and ground into a fine flour. The flour is then sifted to separate the coarser pieces. A little moisture is mixed into the flour to help press it into small, sticky loaves that are baked or roasted until slightly browned. They are eaten as a delightful sweet treat. Common reed flour can also be mixed with wheat and other flours to make breads and cakes.

Common reed is all about sugar. The culms, when wounded, release a sugary sap that oozes out and crystallizes into a gum. First Nation peoples would cut and bundle the culms, and then beat them over a mat to liberate the gum. They would roll the gum into small balls and eat it fresh or toasted. A sweet, licorice-flavored drink can also be made from the gum by simply stirring it into water.

Leaves

The young leaves are edible. They are gathered when they are beginning to unfurl and can be steamed or boiled as a green vegetable or brewed for tea.

The mature leaves and upper portion of the stems can also be eaten. They are boiled down until all the water has evaporated, leaving behind a sticky residue that is rolled into balls for a tasty, sweet treat.

Seeds

The seeds are eaten, but their yield is modest. Various tribal groups burned off the silk that clings to them, then crushed the seeds hulls and all (hulling them can be very difficult if not impossible). After thirty or so minutes of cooking, they are ready to eat in the form of a gruel. Common reed seeds are high in fiber and rate nutritionally between wheat and rice. The most popular way to eat the gruel is to mix in berries and some sort of sweetener.

Salt and Trace Minerals

Some tribal groups in California extracted salt from common reed if it grew in coastal salt marshes. Plant salts tend to be considerably richer in essential trace elements than the refined salt found in grocery stores. Of particular importance is their iodine content. Common reed grows abundantly around the Great Salt Lake in Utah.

Medicine:

Perhaps nowhere else on earth has the use of Phragmites for folk medicine been better documented than in China. This dates back at least to the 16th century when it was discussed in the *Pen Ts'ao* (Materia Medica) by the scholar Li Shih-Chen. The principal parts used medicinally are the rhizomes (lu-ken) and the culms (wei-ching). However, the new shoots, leaves, flowers, and ash are each used at times for specific conditions. In TCM, the earliest form of scientific medicine, the rhizomes and culms were used as an antipyretic, refrigerant, and demulcent. They can soothe irritated and inflamed tissues, particularly in the lungs or stomach.

Wherever illness is connected to excess internal heat, such as fever, common reed is applied as a healing agent. It is also thought to function as a normalizing stimulant to the body's fluid secretions. For instance, it has been used as a diuretic to treat cases that involve painful urination (dysuria), blood in the urine associated with kidney disease (hematuria), or suppressed urine flow.

When excessive stomach heat causes nausea and vomiting, it can normalize the condition. It can be used to increase the flow of saliva when dry mouth or thirst from exertion occurs. It can also be used for treating foul sores or tumors, particularly where blood, pus, or lymph fluids have accumulated. It has been used to treat abscesses, condyloma, generalized edema, and indurations. For typhoid and other fevers, common reed promotes copious perspiration which brings down temperatures through evaporative cooling.

For irritations of the mucous membranes, common reed acts as a soothing demulcent and is said to be helpful for arthritis, gout, and rheumatic conditions in general. It is used as an antidote for food poisoning (particularly from seafood) and has been used in folk medicine to treat leukemia, bronchitis, coughing in general, excessive phlegm, various types of cancers (including breast cancer), jaundice, and optic alexia. It is also used to stem internal and external hemorrhaging, and is thought to help normalize symptoms of cholera, diabetes, flux, and even hiccups.

Other Uses:

Few if any plants have been so intimately intertwined with human culture and history. From ancient times to the present, common reed has played an essential role in the economies of small tribal groups and great civilizations alike. It was used for alcohol production, arrow shafts, basket weaving, brooms, compost, cordage, cradle backing, crafts, duck traps, dice, dye, drying racks, fishing poles, flutes, games, insulation, lattice, livestock fodder, loom rods, mats, measuring rods, nets, pipe stems, poles for harvesting fruit, prayer sticks, reinforcement for adobe structures, reed fishing boats, screens, snares, string, thatch, trellising, wattle fencing, and whistles. Many of these articles are still being made and used today.

In contemporary times use of common reed has grown to include blinds, cardboard, cellophane, clarinet reeds, cork substitutes, fiber board, floral arrangements, paper pulp, pipes for organs, a source of rayon and various other textile fibers, and windbreaks. It is also used for erosion control, as a soil conditioner, and in sewage and gray water treatment.

Common reed played a significant role in two exceptional Empires, the Mesopotamian and the Incan. Mesopotamia was located in present-day Iraq, where even today common reed is essential to the Marshland Arabs, descedants of the Mesopotamians. In addition to many of the uses already described, the Marshland Arabs build their homes out of common reed and may even have placed these homes on reed rafts that could move about on rivers, side-channels, and lakes, essentially functioning as the first mobile homes.

The Incans did something similar in Lake Titicaca. At an altitude of 12,500 feet and an area of 3,500 square miles, it is the world's highest navigable lake. Today you can find descendents of the Incan Empire living on man-made islands constructed from common reed, in houses made of common reed, and fishing in boats crafted from common reed. Indeed, one might find the families that live on these islands having a meal of tender reed sprouts and fish. Today there are roughly 150 of these islands on the lake.

Common reed was also an important economic resource for First Nation peoples of the Southwest. The Puebloans ate the roots and used the reeds for thatch, roof insulation, basketry, cordage, cradle backs, dice, mats, nets, prayer sticks, and screens. The Hopi used it to make flutes, pipe stems, and weaving rods. The Zuni packed the hollow stems with tobacco to smoke in rituals or while praying. The Cheyenne of the western plains used it medicinally, as did the Navajo.

At 1,700 miles long, the Danube is the second longest river in Europe. The last leg of its journey to the Black Sea takes it through Romania. Huge colonies of common reed, some forming floating fens, are found where the Danube meets the sea. In the 1960s and 1970s Romania had a rapid industrial expansion. Looking at the abundance of common reed in their country as a natural resource, the Romanians began using it to manufacture cardboard, cellophane, fiberboard, insulation, and textile fiber.

In Phragmites we see a plant that has much to contribute to a nature-based economy in any region it grows naturally. Many small-to-medium-sized enterprises can benefit from the numerous products that can be derived from it. And yet today in the United States the ecological and economic importance of common reed is being ignored. It has also been the victim of human development. Paved over wetlands, dammed rivers, and diverted water for irrigation has done away with common reed habitat. To make matters worse, it has been declared a weed due to its aggressive growth in hydrophilic areas by misguided environmentalists who attempt to eradicate it with toxic herbicides.

Renewable Energy

Common reed can yield up to 25 tons of renewable biomass per acre. Its biomass could be used to fuel a pyrolysis kiln, in conjunction with a boiler, to create steam to power a generator and produce electricity. The steam is just a waste gas and the electricity just a byproduct. The real end product is the resulting charcoal called agrichar or biochar—a high quality agricultural soil amendment that sequesters 85 to 95% of its original carbon when returned to the soil. Other byproducts of the process are steam heat, humidification, medicine, fertilizer, and a termite repellent, all with near zero air pollution.

Ethanol and Solvents

Common reed is often fermented for alcohol production and can be used for biofuel. It is used as a solvent for gums, resins, lacquers, and varnishes, as well as for essential oils used in perfume, tinctures, and other pharmaceutical purposes. The alcohol can also be used in the manufacture of dyes.

Sewage Treatment

Common reed's ability to absorb toxins and other contaminants from water make it ideal for natural sewage treatment or as a tool to clean up pollution. In addition, it can dry sludge and flocculate extremely small particles.

Ecological Functions:

Common reed is often found in wetlands and riparian corridors throughout much of the world where they contribute significantly to cleansing pollution, controlling floods, and sheltering wildlife. Wetlands are perhaps the most productive ecosystems on land. For example, oxygen production is greater acre-to-acre in wetlands than in many virgin forests, which are already a valuable source of oxygen themselves. Wetlands, in proportion to their size, offer valuable and abundant ecological and economic functions and services. They act as critical bio-filtration systems that capture pollutants like excess nitrogen, phosphorous, sulfates, copper, iron, and other heavy metals. Aquatic flora cleans up these toxins by consuming them, and then depositing a great deal of them beneath the water in the anaerobic muck where they become trapped. Another valuable asset of wetlands is that they maintain a balanced pH in the water.

Wetlands also help recharge aquifers by slowing the mountain runoff of rain and snowmelt so that it can be drawn into the soil and reduce the likelihood of floods.

Wetlands play an essential role in the lives of many habitat dependent species, such as geese, ducks, cormorants, swans, storks, herons, egrets, gulls, terns, and other water fowl. Migratory waterfowl need an unbroken chain of wetlands across their migratory paths to maintain their cyclical lifestyle. Wetlands also offer critical habitat to many mammals, fish, and insects. The common reed is a critical part of all of the wetlands' value.

Native Range and Habitat:

Common reed is native to every continent except Antarctica, and can also be found on many islands. In the Americas it grows across Canada from British Columbia to Nova Scotia. In the United States it grows coast-to-coast and in every state except for a large area in the inland Southeast. Common reed extends down through Mexico nearly to the southern tip of South America. It is also commonly found in Europe, the Middle East, Asia, Africa, and Australia. Although common reed is most prevalent in coastal areas, it is in no way limited to them. Common reed is a wetland plant found growing along rivers, streams, lakes, ponds, springs, bogs, salt marshes, seeps, alkali sinks, and roadside swales from sea level up into the mountains.

This incredibly ancient member of the grass family has an extremely diverse number of associate species, ranging from cattails to wild rice and greasewood to creosote bush.

PROSOPIS SPP.
(Mesquite)
[Leguminosae Family]

Mesquites are members of the Leguminosae family, and the sub-family of Mimosoideae. There are twenty-five to forty species worldwide. Most are found in South America; a few are natives of India and Africa. In the southwestern United States, there are at least two species. One of these species, Prosopis glandulosa, has repeatedly confused botanists, and has been labeled under various names, often erroneously. Minor differences would suggest that some, if not all, of these are subspecies of Prosopis glandulosa. Their biggest distinctions, however, seem to be geographical, since they range as far east as Kansas, through Oklahoma and Texas, west to Southern California and south to Mexico.

Prosopis Glandulosa
[Syn. P. Chilensis var. Glandulosa, or P. Julifora var. Glandulosa]
(Honey Mesquite, Mizquitl)

Honey mesquite includes sub spp. P.G. var glandulosa, P.G. var. Torreyana [syn. P. Juliflora sub spp. Torreyana], and P.G. var. Velutina [syn. P. Juliflora sub spp. Velutina](Velvet Pod Mesquite). Its name is derived from the Aztecan "mizquitl." It is called "algarroba" by Mexican Americans and "ily" by California's Cahuilla peoples, or "ah pe" by the Paiutes.

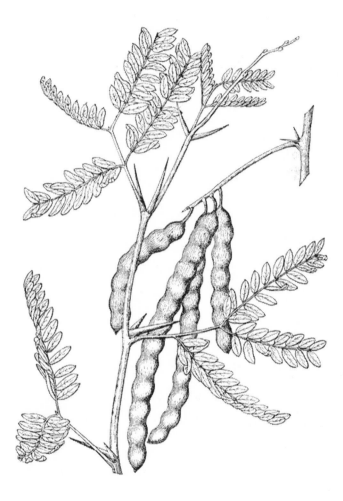

Prosopis Glandulosa

Mesquite is slow growing in poor dry soil, but grows fairly rapidly in good soil with consistently available moisture. It is a long-lived, deciduous tree or shrub, usually 10 to 20 feet tall. In flood plains and riparian zones with high water tables, it can grow faster and can reach heights of 40 to 60 feet at maturity, with open canopies 40 to 70 feet across. Most of these big trees, however, were cut down long ago. Once trunks 2 to 4 feet in diameter were known, but most mesquite trunks today are short and only 6 inches to 1 foot in diameter. The bark is thick and rough, tending to peel with age. Bark may be ash grey to dark brown, often with reddish highlights. The branches are mostly thorny and crooked, and tend to droop. The honey mesquite blooms from May to June but can be found blooming anytime between spring and September, attracting droves of small insects. The flowers are mildly fragrant and form slender cylindrical spikes 1½ to 4 inches long that are yellowish or creamy green in color. The flowers develop into clusters of brown seed pods, that are sometimes a rich mahogany. The pods are 2 to 9 inches long, about ¼ to ½ inch thick, and constricted around the seeds. They are usually straight, but occasionally curve. In the spaces between the seeds a thick sweet pulp is found. Pods mature from late August through October. The foliage color varies with location and among individuals, ranging from bright green to dark green, but may be yellowish green or even a dull gray green. The foliage is composed of numerous small leaflets that give honey mesquite a ferny look. The branches and the reddish brown twigs are often well-armed with stout straight thorns, ¼ to 1¼ inches long. Young trees are usually spinier, and thorns decrease with age. Occasionally a thornless individual occurs. These might make a good choice for developing cultivars, and could be bred with high yielding individuals.

Mesquite is reputed to have the most extensive root system of any plant. Horizontal roots spread out from the trunk 50 feet or more in all directions. The tap root is huge, and more wood is often found below the surface than above. It is not uncommon for tap roots to dive down 30 to 80 feet, and in the extreme, may reach 150- to 200-foot depths. Roots grow much faster than the above ground plant.

In the pre-settlement days, mesquite forests with large trees were common in drainages throughout the Southwest, with the trees often spaced 40 to 100 feet apart in these wild stands. These trees, however, were harvested for their wood centuries ago. Today, cattle have brought mesquite up from its natural habitat to the prairies, where it typically forms dense thickets. When mesquites are cut down, they sprout into low-growing brush, making it easy for cattle to eat the pods. Large, old trees in their natural habitat are rare today, while the prairie's shrubby mesquites have invaded with a vengeance.

Culture:

Mesquite is sun loving and resistant to reflected sun. It is most common on sandy soils, but does well on most well-drained soils. It will grow in heavy clay soils, but with less vigor, and is adapted to saline and alkaline soil. Mesquite is extremely drought tolerant, reputed to survive on a scant 3 inches of annual precipitation, but it grows faster and bigger with a high water table or occasional deep watering during dry periods. They are hardy to about 0° F and may sprout back even if killed to the ground by subzero temperatures.

Water mesquites in heavy soil modestly, if at all, but they will need regular watering during the first year to get established. Mesquites endure high heat very well, and can grow under adverse conditions and resist wide temperature fluctuations. Mesquite responds well to pruning, but watch those thorns, and be aware that the roots can be invasive.

When planting, container grown plants or bare root are both fine. Propagate by root division or by fresh (not dried) scarified seed. Germination is most successful when soil and air is warm. Cuttings typically fail.

Care must be taken not to damage roots at planting time to assure successful establishment. To prevent blow down, stake larger plants their first year in the ground, then be sure to remove the stakes.

Honey mesquites are remarkably free of most serious pests and diseases, however, they are susceptible to Texas root rot.

Its most serious pest is the mesquite girdler (Oncideres pustulatus), a beetle that burrows under the bark and kills the tip shoots, and sometimes whole branches. Once the mesquite girdler has done its damage, another beetle (Megacyllene antennatus) usually attacks the resulting dead wood, rendering it soft and punky, useless for most purposes.

Bruchid beetles lay their eggs on the surface of honey mesquite pods. When the larvae hatch, they burrow into the pod and eat the seeds. Some bruchid species' larvae feed on the pulp as well. Mesquite, at times, may also host mistletoe, scale, or thrips. At the seedling stage, mesquites often fall victim to hungry rabbits if not protected.

In tropical countries, honey mesquite can become a serious weed, as it has in Hawaii, the Philippines, and the West Indies.

Food:
Pods
The pods are edible, sweet, and tasty, but depending on the stage at which they are collected, may be slightly astringent. High in sucrose (17.5 to 35%) when dried, the pods are also a

rich source of calcium, magnesium, zinc, potassium, and iron, plus fiber and carbohydrates. The pods are typically richer in protein than corn, with less fat and fewer carbohydrates. Protein content (13 to 30%) is similar to beans and rice. Four tablespoons of ground mesquite pods supply about 70 calories. They are also a rich source of lysine, an amino acid that aids in the digestion of protein and is used by the body for growth and tissue repair.

The pods are collected at different stages of growth, depending on the desired taste or use. Some pods may be infested with beetle larva. These can be separated, although native people simply ate them larvae and all, no doubt benefiting from the additional minerals. After harvesting, the pods are usually sun dried, then ground into a meal called pinole or into coarse flour. Using water to moisten the pinole, it was traditionally shaped by hand into small sweet cakes or rolls, and eaten as is. It can also be cooked with soups, stews, and gravies, acting as a thickener and flavoring. One tribe employed solar cooking by packing the pinole firmly into a tightly woven basket, burying it partially in the hot desert sand, and allowing it to bake until firm—often many hours.

Once dried, the pods can be put up for storage for future use. According to Walter Ebeling in his book *Handbook of Indian Foods and Fibers of Arid America,* one interesting low tech grinder made in prehistoric times, known today as a "gyratory crusher," could grind large quantities of pods quickly and efficiently, both crushing and grating simultaneously. Today a hammer mill makes an effective substitute for the more ancient method.

The fresh pods are also used for a beverage called atole. First the pods are crushed, then mixed in boiling water, steeped, and finally cooled and strained. An alcoholic drink can also be made by fermenting this liquid to produce a weak beer. The seeds (beans) are also occasionally used alone to prepare a beverage.

Tender immature pods can be eaten fresh, steamed, boiled, or even added to stir fries. In some cases the seeds are removed by winnowing and used alone for pinole or to make flour for baked goods. The mesquite beans can be baked after soaking in water for 8 to 12 hours. Some people would remove the seeds from the pods and scrape out and eat the pulp by itself.

In recent times, mesquite pods have been processed to make a crude molasses-like sugar substitute which is then made into mesquite candy. The pods have also been made into pudding or used as a flavoring for corn chips or broth. Being a legume, perhaps mesquite tofu or tempeh is not far off.

A healthy, older mesquite on deep, moist, rich alluvial soils can yield 50 to 150 pounds of pods each year. Smaller trees may produce one to several bushels at about 8 gallons per bushel. In good years, two crops can be harvested, yielding up to 1,000 pounds per acre. One hundred bushels per acre is fairly common. Thickets can be thinned to fewer, larger plants to increase pod production.

Dry pods that have fallen to the ground beneath mesquites are quite edible in dry periods, and are said to be highest in carbohydrates at this stage. However, during the rainy season, fallen pods quickly spoil and become inedible.

Inner bark
The inner bark can be used as a substitute for rennet to curdle milk for making cheese.

Sap (Gum)
Mesquite sap is also a nutritious food source as a gum. It can

be eaten as is, but more commonly has been made into candy, much like gum drops. The gum oozes from the bark of wounded or stressed mesquites or can be induced by cutting branches from the trunk. About three or four weeks afterward, the gum is harvested from the wounds.

Other Food Uses

Mesquite flowers can be roasted and eaten, or used for herb tea.

By combining mesquite with a few of its eco-associates, such as live oaks, prickly pears, and yuccas, a first class dry farmed polyculture could be developed for arid and semi-arid land. The live oaks' acorns are high in beta carotene, while mesquites are low; live oaks are low in sugar, while mesquites are high.

Prickly pears, yuccas, live oaks, and mesquites, when combined, provide ample digestible calories. All four plants supply fine-tasting, very nutritious foods, and all but the yucca are potentially high yield food crops. The yucca, however, offers numerous other economical products (see page 232).

Medicine:

Mesquite has been valued since prehistoric times by the Aztecs and other tribes of Mexico and the southwestern United States for its medicinal properties. Spanish settlers quickly learned to value mesquite's healing virtues and adopted it into their folk medicine. The gum, leaves, flowers, pods, inner and outer barks have all been used by native peoples for medicine. It is very astringent, antispasmodic, and moderately antimicrobial.

The infused leaves or pods were used to treat a variety of eye problems, such as inflammation, irritation, and pinkeye. The gum is insecticidal, and an effective remedy for head lice.

Other external applications employ mesquite as a salve, poultice, or as a wash for sores, scrapes, cuts, chapped or split lips or fingers, sunburns, or inflamed hemorrhoids.

On the glycemic index, mesquite meal registers quite low at 25, making it a good choice for the diets of diabetics. Herbalists have suggested internal use of mesquite for various forms of congestion, particularly of the lungs, as well as for sore throats and laryngitis. Other internal treatments include inflammations of the colon and bladder, excess stomach acidity, chronic indigestion, diarrhea, peptic ulcers, umbilical hernias, fever, acute pain, and as nourishment for newborn infants. For an excellent medicinal profile of mesquite see Michael Moore's book *Medicinal Plants of the Desert and Canyon West*.

Agricultural Uses:

Bee Forage

Mesquite is one of the most important bee forage plants in the Southwest, producing more honey than the majority of bee plants in the arid and semi-arid regions of the United States. A single colony can produce 80 pounds of mesquite honey per season, and in Texas there have been reports of up to 200 pounds per season.

Mesquite honey is amber colored and has an excellent flavor, although it can quickly cloud and turn grainy. In good years, mesquites may bloom twice, and a single plant could produce as many as 12 million flowers, supplying both wild bees and honey bees with vast quantities of nectar and pollen. In return, they pollinate the mesquites.

Various environmental factors conspire against mesquite's incredible potential as a honey plant. Late frosts in the northern section of its range, or at higher altitudes, can kill it back, sometimes to the ground. Mesquites are tough, however, and usually resprout quickly, although the flower crop in such times is greatly reduced for the year. For this reason, in the coldest part of the mesquite's range, high yields of honey may only occur in 2- to 4-year cycles.

Other yield restrictions include below average spring precipitation, followed by an unseasonably cool summer. Mesquites are also least productive when they grow on heavy clay soils. That said, when used on an acre-to-acre basis as a bee forage, mesquite honey production is often more profitable than raising cattle.

Beneficial Insect Habitat

Like bees, beneficial insects use mesquites as habitat because of the abundance of protein and carbohydrates their flowers offer. In addition to serving as pollinators, they act either as predators or parasites to control agricultural pests.

Soil Enrichment

Mesquites are top producers of nitrogen for soil enrichment among arid land plants. The foliage contains about 4% nitrogen, and the pods are nearly as well endowed. Because of the high quality soil they generate and the light shade cast by their open canopies, many low-growing plants thrive beneath them.

Livestock Forage

Mesquites have excellent forage value for all classes of livestock, and are resistant to heavy grazing. The pods have a high feed value and can quickly put weight on livestock. The animals love the pods and eat them greedily. Some livestock, cattle for instance, have difficulty digesting large quantities of pods, while others, such as sheep, do not. Goats and hogs, as well, consume the pods with equal fervor.

In the early 1950s, Professor Russell Smith reported in his landmark book *Tree Crops, A Permanent Agriculture* that a commercial oil mill was successfully employed to press ground pods into large cakes. The cakes from the initial pressing were used for the next two years without spoiling to feed livestock. Milling the pods and adding roughage make the pods significantly more digestible for livestock. The leaves of mesquite are rich in fiber, 16 to 20%, however, livestock typically ignore the leaves if more desirable forage is present.

Livestock Containment
Two functions can be served when the thorny mesquites are used as livestock barriers. Fence them in or out, whatever the case may be, and let them feed off the densely planted honey mesquites. Native people would often grow their gardens using mesquites as protective enclosures against wildlife or livestock intrusion. The nitrogen-fixing mesquite makes the soil fertile, while also providing the garden protection against damaging winds and blowing sand.

Livestock Associate
It makes sense to raise selected livestock, such as wild turkeys or possibly small herds of Churro sheep or dwarf wool bearing goats, in among the mesquite.

Other Uses:
Wood
Honey mesquite wood is a deep, rich brown or reddish brown, with light yellow sapwood. The wood is nicely patterned, close grained, and takes a fine polish. It cures well without checking (cracking) and is very rot resistant when in contact with the soil. It has value for post and beam construction, cabinetry, flooring, paneling, molding, veneers,

furniture, tiles, frames, tool handles, various kinds of utensils, miscellaneous wood carvings, and other woodwork.

The National Wood Flooring Association (NWFA) rates mesquite as an outstanding commercial wood, and it is commonly used for hardwood floors. As mesquite wood ages it turns a rich, rose color similar to cherry. Compared to red oak, it is almost twice as hard and has nearly 350% greater dimensional stability, meaning less shrinkage and swelling due to fluctuations in humidity. Mesquite hardwood floors have an insulation value equal to 15 inches of concrete.

Mesquite woodlots often harvest the wood when the plant reaches 6 feet in height. In a harsh, droughty, arid environment this could take twenty-five years. With good soil and plenty of water, mesquite can reach six feet in a fraction of that time. If the whole plant is cut to the ground at harvest, which can be done year after year, it will root sprout and grow back even faster. Its ability to grow back after cutting makes mesquite a valuable renewable wood resource. Thickets can be thinned to fewer, larger trees to increase wood volume, length, and stem diameter. Annual yields of seven tons per acre would not be unusual.

The wood is useful for firewood and is a good biomass producer. It is very hard and heavy; it burns hot and is as good or better at BTU production than many other firewoods. Mesquite wood burns for a long time, and the coals are very long lasting. The delightful fragrance while burning accounts for some of its popularity as a fuel wood, particularly for smoked foods and barbecue. It has also been used as an excellent source of fuel for firing pottery.

In recent times, it has been widely marketed as a charcoal cooking fuel, as in barbecuing, but it has been disappointing because the lovely aroma is consumed in the production process. The loss of mesquite's rich habitat, important to so many diverse species, especially along drainages, makes mesquite charcoal fuel production questionable ecologically, as is ripping them out by the roots or killing them with massive doses of herbicides, to which mesquites have proven to be highly resistant.

Tribal people used the wood to construct their homes and for many other purposes, including the making of hunting bows and digging sticks, which were used to lift edible and medicinal roots and bulbs from the ground.

At one time paving blocks were made from mesquite. In San Antonio, Texas, for instance, several streets were once paved with mesquite blocks. They held up very well for more than twenty-five years.

Mesquite chaparral produces a minimum amount of above ground wood, the exception being in riparian corridors where good soil and plenty of water are available. In general, however, a greater volume of wood is produced below ground then above, and it is barely worth the labor it takes to unearth it. Once, however, it was not uncommon to use a team of horses to yank mesquite roots out of the ground for the wood. More recently, large machines designed for this purpose are being used in an aggressive effort to restore the prairie to grassland for the sake of cattle.

The goal of restoration is to increase the grasslands so larger numbers of cattle can be stocked. The irony is that overgrazing, principally by cattle, is responsible for mesquite's invasion of the prairie in the first place.

Ethanol
Mesquite pods contain between 17.5 to 35% sucrose, making them a prime candidate for ethanol production. In David Blume's book *Alcohol Can Be a Gas*, he says that as much as 341 gallons of ethanol can be produced per acre annually

from mesquite pods. Blume estimates if all the wild mesquites were harvested for biofuel, their pods could yield as much as 23.9 million gallons of ethanol annually.

Sap (Gum)

In addition to its edible and medicinal qualities, the clear gum that exudes from mesquite works well as an adhesive, and was used, for example, to mend broken pottery. The gum has also served as an ingredient in laundry soaps and been used to produce a black hair dye that adds luster to the hair. The gum can be boiled and used as black paint to embellish pottery or as a varnish. Native people used the gum in cosmetics.

Tanning

The bark, which is high in tannic acid, has been used by tribal groups to tan leather

Fiber

A coarse fiber can be liberated from both the inner and outer tree bark, as well as the root bark. Fine basketry has long been produced utilizing this fiber. It was also used to weave a coarse cloth that was beaten and manipulated until soft, and then used for making women's skirts and baby diapers.

Dye

A blue dye can be extracted from both the leaves and the pods.

Wind Break

Mesquites are exceptionally wind firm because of their deep, thick taproot and can be used for wind breaks. The draw back is the lack of foliage density, rendering them quite porous.

Erosion Control
Mesquites possess an amazing ability to control soil erosion due to their exceptionally large root system, and are often planted to stabilize sand dunes. While capturing blowing sand, the mesquites themselves are often partially buried and hummocks form around them—yet they survive and continue growing unperturbed.

Ecological Functions:
Arid Ecosystems
In its natural habitat of the Southwest's river drainages, honey mesquite communities, with their diverse plant and animal species associates, are considered the most productive arid ecosystems in the world.

Wildlife Habitat
Numerous birds and animals depend on honey mesquites for habitat (shelter) and niche (resources). Deer, coyotes, rabbits, roadrunners, skunks, javelina, squirrels, rodents, lizards, horny toads, king snakes, rattlesnakes, and coral snakes find abundant food sources and excellent cover in mesquite bosques. Quail, roadrunners, and doves are just three of the many bird species that live among the mesquite; indeed, the shredding bark is excellent for building nests.

Native Range and Habitat:
Honey mesquite is native to California, Arizona, New Mexico, Texas, Baja California, and Mexico's Sonoran Desert. It has naturalized in Nevada, Utah, Colorado, Oklahoma, Kansas, Louisiana, Arkansas, and now covers close to 82 million acres of former Texas prairie. In Utah, honey mesquite is found only in the southwest corner. In Colorado, only a tiny

population exists in the southeast on Mesa de Maya. It is common in the Mojave and Colorado deserts. It was once abundant in the lower Colorado, Rio Grande, and Gila River drainages.

Once limited to locations with high water tables—flood plains, river and stream drainages, canyons, desert washes, valley basins, salt flats, and arroyos—honey mesquite now grows at sea level and at altitudes up to 6,000 feet. Though more common on prairie grasslands, it can also be seen on hillsides, mesas, and dry alkali sinks.

At lower elevations, honey mesquite often integrates with creosote bush. At the highest altitudes, it can be found with piñon/junipers. Other mesquite associates include live oaks, acacias, hackberries, desert willows, jojobas, saltbushes, yuccas, prickly pears, screwbean mesquites, various cacti, and sand verbena. Next door neighbors in the upland riparian zones include cottonwoods, willows, soapberries, walnuts, hackberries, sycamores, wild grapes, Clematis, and various short grasses.

Prosopis Pubescens
(Screwbean Mesquite)

Known to Mexicans, Mexican Americans, and southwestern Spaniards as "tornillo," the Paiute people call Prosopis pubescens "quier." Screwbean mesquite is a slow growing, wide spreading, deciduous shrub or small tree, reaching 10 to 20 feet in height. Its trunks are up to 1 foot in diameter, with grayish brown bark that peels on older plants. In general, screwbean mesquite is smaller than the honey mesquite, and is nowhere near as abundant. The screwbean's flowers bloom May through June. They vary from cream colored to bright yellow, are borne in 2 to 3 inch spikes, and are produced in high number. The leaflets and the ferny effect they impart is much like honey mesquite's, except the leaflets are smaller, and their twigs are gray rather than reddish brown. Although well armed, their thorns are smaller than those of honey mesquite. As their name suggests, screwbeans produce very unique seed pods. Each pod is shaped in a tight, buff colored spiral. The pods grow in clusters and are quite distinctive when compared to the bean-like pods of honey mesquite, or indeed with the fruit of any plant. The pods are usually 2 to 3 inches long and ripen July through August.

Culture:

Screwbean mesquite loves the sun, but is tolerant of partial shade. It prefers fairly rich soils but does well on most soils with good drainage, even those that are saline. It is drought tolerant but not to the extent that honey mesquite is.

Food:

Pods and Seeds

The pods and seeds can be used for food in almost all the same ways as honey mesquite. At harvest time, however, the fresh pods are rather bitter, and their flavor is less than pleasant, unlike their cousin. With slow cooking though, they become very sweet, sometimes even more so than honey mesquite. After cooking, they can be dried, and then stored for a year or more.

The pods are exceptionally high in sugar, and although they are an excellent source of protein and fiber, they are less endowed than the honey mesquites. They are often boiled down into a thick, sweet, molasses-like syrup.

Seed pod production is high, and in one documented case, a 15-foot screwbean yielded 1½ bushels of pods.

Unfortunately, they are more likely to be infested with bruchid larvae than the honey mesquites.

Agricultural Uses:
Forage
The forage value and palatability of the screwbean mesquite is high for livestock and wildlife alike, but it is reported that its flowers, unlike honey mesquite's, are toxic to honeybees.

Other Uses:
Wood
The wood of the screwbean is very hard. It is also highly ornamental, even more so than that of honey mesquite. The highly patterned grain possesses light and dark highlights of reddish brown, and the sapwood is a soft yellow color.

Native Range and Habitat:
The screwbean mesquite is native from Southern California to Texas and farther south into Mexico. It usually grows on alluvial soils along river and stream basins, but it is also found on flood plains, canyon walls, bottom land, arroyos, near springs, or in low depressions—and always at lower altitudes.

ROSA SPECIES
(Rose)
[Rosaceae Family]

For at least 3,000 years, the rose has been regarded as the queen of flowers. There are over 100 species worldwide, most being native to the temperate zones of the Northern Hemisphere, with at least seventeen species native to the United States. From this relatively small number of species, it is estimated that about 20,000 cultivars have been recognized.

The rose has come to symbolize love, beauty, pleasure, and fertility, and has long been cultivated in response to such urgings. Numerous far-flung cultures of the old and new worlds have valued roses for food, medicine, beauty, and the intoxicating ambrosia emitted by the flowers of some species. For good reason, the bond between roses and people is ancient.

Roses in General
Culture:
The species of roses profiled here are easily grown. In all cases, they are sun lovers that prefer eight hours of full sun daily, but they can do nearly as well with some late afternoon shade. In half shade, they may grow well enough, but flower and hip production will be much reduced.

Soil preferences vary to some extent, but they all adapt well to most soils. They favor good drainage and clay loam, but tend to do well in heavy clay, if it is not soggy. They like soils rich in organic matter, yet some may flourish in pure sand. They favor a pH near neutral, and seem content with a

Rosa Multiflora

pH that is either slightly acid or slightly alkaline. Very wet soils and standing water will injure or kill them.

Roses are subject to numerous pests and diseases with some exceptions, such as the rugosas. As a preventative measure, always plant them where they will receive good air circulation.

Some common rose diseases include black spot, powdery mildew, rust, and various viral and canker diseases. Common pests include aphids, leafhoppers, thrips, Japanese beetles, spider mites, scale, rose bugs, rose slugs, and borers.

Work granulated kelp meal into the soil as a fungal disease preventative, and use liquid kelp concentrate and/or garlic spray as a treatment. Do not use nitrogen fertilizers, but rather keep roses side dressed with composted manure.

Many gardeners in the United States and Europe believe that garlic and parsley are beneficial associates for roses and will plant guilds of the three together. However, there is no research to support literally centuries of observation.

Parsley is an excellent habitat for beneficial predatory insects. Garlic is both an antibacterial and a fungicide, as are onions and chives. Garlic sprays applied to roses have successfully repelled attacks by various pests, including at least one species of aphid for up to thirty days from a single application.

Pruning is essential for good floral productivity. It is also needed if an incredible snarl is to be avoided. A modest annual pruning is all that is required, but skip a year or more and the job can reach nightmarish proportions. Prune in the spring, after the danger of a late frost is minimal, and when the buds begin to swell on the stems. Remove dead branches and cut out old unproductive stems to encourage flowering. Otherwise prune to inhibit unwanted overgrowth and tangled branches. While Rosa multiflora may need a great deal of pruning, Rosa rugosa will need little to none. Most species are easily propagated by suckers or stem cuttings from young, flexible canes following flowering. (For more on culture see the individual species.)

Food:

Rose Hips

Rose hips (the fruit) are usually dried, cooked, or pureed before being used to make jellies, syrups, soups, sauces, teas, sweets, wine, or pie filling—or added to puddings, breads, muffins, and tarts. Corn starch or potato flour are sometimes added to hip jams and jellies to temper their astringency.

For tea, two tablespoons of fresh or dried hips are used per pint of water. The water is heated, but not boiled. The hips are added, covered and steeped for about twenty minutes, then strained and served. In this way the highly perishable vitamin C is preserved. It also helps to first chill the hips to inactivate the vitamin C-destroying enzymes.

Rose Petals

The flower petals are not only delicious to eat, but can also sweeten the breath. All species of rose petals are edible, but only the really fragrant ones are very tasty. They are mixed into salads, used as a garnish, or added to rose petal jam, syrup, or vinegar. Petals can also be dipped in batter and quick fried. They are popular when candied with crystallized sugar and eaten as a sweet treat or used to decorate cakes. The petals can also be added to sandwiches, or even used to flavor ice cream. The flower buds of one species, Rosa centifolia, are pickled, while in other species, like Rosa multiflora, they are parboiled before being eaten.

Rose Water

In a number of cultures, rose water is used as a food flavoring, particularly for sweets. Arabic, Iranian, and East Indian cuisine all use rose water in the preparation of certain traditional foods. Rose water, in addition to flavor, supplies meals with its delicate fragrance.

Essential Oil

The essential oil of roses is distilled commercially from the flower petals. Cultivation for oil began in the 16th century and continues today. The oil is commonly known in the trade as "Attar of Roses," although there are also regional names, such as "Essence of Rose of the Balkans," or it may simply be called "Oil of Roses." Care must be taken with this last name to make sure it is the essential oil and not vegetable oil that rose petals have been soaked in.

Essential oil of roses is extremely expensive because such a huge quantity of rose petals is needed to distill a very modest amount of oil. The flower petal to oil ratio will vary by species, culture, and climate, but to yield a mere ounce of oil, it can take two tons of petals—that is probably about 60,000 flowers. The resulting essential oil is a delicate, greenish yellow to orange color and deeply aromatic.

As flavoring, essential rose oil might be found in baked goods, beverages, fruit desserts, sweet treats, and ice cream.

To produce the essential oil of rose, fresh petals are collected and processed immediately. Some believe the best flowers are those collected on mornings when dew has formed. In Bach's flower remedies, it is the dew itself that is most valued. Bach's Rose Remedy is prescribed for apathy, passivity, and lack of motivation. For fatalists who have lost interest in life and surrendered themselves without resistance, this remedy is intended to induce an appreciation of life.

Leaves

The leaves of some rose species are aromatic, and their young leaves may be brewed for tea or used as a seasoning, either fresh or dried.

Medicine:

Rose Hips

Vitamin C is an important antioxidant that strengthens blood vessels, inhibits premature aging, prevents strokes, reduces hair loss, and builds healthy teeth and gums. Fresh rose hips are extremely rich in vitamin C, reputedly fifty to sixty times more endowed than lemons and up to twenty-four times more than oranges. However, their species, culture, harvest periods and practices, drying methods, and storage can effect the amount of vitamin C present. Dried hips in particular can lose much of their vitamin C content if they are not handled properly. Unfortunately, it is sometimes true that dried hips found in the marketplace have very little of this important antioxidant remaining. Similarly, the constituents and medicinal effectiveness of rose hips vary between species.

The rugosa rose is generally the species of choice for hip production, due to its unusually high vitamin C content and the large size of the fruits. Rugosas are cultivated commercially for the manufacture of natural vitamin C supplements. The hips are harvested in late summer or early fall, usually before the first frost.

Rose hips possess bioflavonoids as well, which enhance the absorption of vitamin C and increase its potency. These constituents also strengthen connective tissue and reduce capillary fragility.

Rose hips are also a good source of vitamins A and E, valuable antioxidants whose presence adds critical balance to the vitamin C content. Also found in the hips are niacin, riboflavin, vitamin K, various minerals, citric, nicotinic, and malic acids. Malic acid, in particular, protects the stomach lining and so helps prevent ulcers.

Medicinally the astringent nature of the hips can shrink inflamed tissues, as well as reduce the secretion of mucous. The hips appear to be strongly antiviral, antibacterial, and also diuretic. Herbalists past and present have prescribed them for chest complaints, influenza, colds, and coughs. They have been used as a depurative tonic to detoxify the blood and internal organs, reduce capillary fragility, and reverse varicose veins. Rose hips are mildly laxative and can bring relief from digestive spasms, constipation, and nausea. Other applications include relief from minor infections, fever, skin inflammations, bruises, hemorrhoids, kidney stones, and general debility.

In parts of Europe, rose hip syrup is taken daily—about a tablespoonful—as a preventative boost for the immune system. The syrup preserves the medical benefits of the hips, increasing the longevity of their active constituents. Otherwise, tea is used.

Rose Petals

Like rose hips, the flower petals are also valued in folk medicine. Indeed they are rich in medicinal constituents, including the bioflavonoid quercetin, which serves as an antioxidant, an anti-inflammatory, and an antihistamine. Quercetin, like other bioflavonoids, increases vitamin C utilization by the body.

The petals, brewed as tea, have been prescribed for colds, stomach troubles like gastritis and diarrhea, depression and lethargy, and for inflammation of the mucous membranes, particularly of the nose and throat. An infusion of the flower petals is gargled for mouth sores and sore throats. The flowers are infused in water for a fragrant vaginal douche.

Flowers should be harvested on a dry day before they are fully open and after all dew has evaporated. The floral base and the stamens are removed so that just the petals remain. If not used fresh, they are then dried. Some herbalists believe the dried petals are more potent than the fresh. Sun-dried petals tend to loss some of their the color, fragrance, and medical compounds, so shade dried petals, spread out on screens, are preferable. They can also be oven dried on low heat for a short duration. Note they must be thoroughly dry to prevent mildew.

Dried petals are powdered or left whole and stored in an airtight, sealed glass container. Initially, their perfume may increase, and can last years, although their medicinal value will probably be lost after three or four months. One pound of petals equals about nine well-packed cupfuls.

Essential Oil

The constituents in the rose's essential oil number in the hundreds but vary to some extent by species. Citronellol is the chief constituent; others include geraniol, nerol, eugenol, stearopten, and phenylethanol. Bioflavonoids include quercetin, B-carotene, and other pigments. The wonderful fragrance of the flowers originates from trace amounts of volatile oil in the petals. For an essential oil, rose oil is notably non-toxic and can be used undiluted.

As might be expected, the essential oil is esteemed in herbal medicine. Its principle therapeutic actions are astringent, antibacterial, antidepressant, antiseptic, antiviral, choleretic, depurative, emmenagogue, emollient, haemostatic, hepatic, sedative, and stomachic. It is also a tonic for the heart, circulatory system, nervous system, liver, stomach, and ulcers.

Rose oil is thought to have a strong affinity with the female reproductive system. It is helpful for regulating menstruation, soothing post-natal depression, and alleviating PMS. It can have a significant aphrodisiac influence for repressed sexuality, or simply to increase one's pleasure.

The oil is prescribed for aiding digestion, for toning the digestive system by increasing the flow of bile from the liver, for aiding the absorption of fats, for preventing putrefaction, for alkalizing the intestinal tract, and for stimulating the appetite. In addition, it is recommended for toning the vascular system, principally as a blood purifier. The oil has also been used to prevent hemorrhaging.

Rose essential oil benefits the nervous system, bringing soothing relief from stress-induced tension. It helps promote sleep for insomnia sufferers and acts as a mood enhancer to lift the spirits of those feeling the blues.

Highly regarded for eye problems, the oil is used in the treatment of eye inflammations, eye pain, and conjunctivitis. Other therapeutic applications include treating inflammations, headaches, earaches, hay fever, and ulcers of the mouth and throat. Aromatherapists often regard rose oil as a universal rejuvenator.

Rose Water

Somewhere around the 10th century, an inspired individual had the presence of mind to make a decoction of the flower petals, and thus a wonderful, multi-purpose balm known as rose water was born. Rose water is highly beneficial to the skin, being simultaneously soothing and invigorating.

Medicinally, rose water is an astringent and an antiseptic. Herbalists may prescribe it as a topical treatment for minor cuts, bruises, sprains, or pulled ligaments. Several cups of rose water tea taken internally daily might be recommended as a tonic for the liver or for nervous conditions or stomach ulcers. Commercial eyewashes commonly include rose water because of its ability to soothe sore or runny eyes. For ulcers of the mouth or throat, rose water is gargled.

Rose water originally was produced from scratch, and still is, however, most of it is generated today as a by-product of the distillation of the essential oil.

There are variations on how rose water is made, but the way that best preserves the nutritional, medicinal, and aromatic qualities is as follows: Heat water at a low temperature, add flowers, let them simmer for a few minutes, turn off the heat, steep for a couple more hours, then strain. Some suggest bruising or crushing flowers first. Use about one heaping handful of petals per quart of water.

Seeds
Rose seeds have been used as a sedative and a diuretic. They are a rich source of vitamin E and contain at least one detergent saponin that is likely responsible for the seeds' diuretic properties.

Roots
A decoction of the roots of various rose species has been used medicinally, particularly as an astringent.

Agricultural Uses:
Honeybee Forage
Roses appear to have an abundance of pollen, and for this reason they provide a valuable service when pollen-bearing plants are scarce. Roses that yield nectar, however, are much less common. Among those that do provide nectar some may only do so every few years. Roses that regularly serve up pollen and nectar yield a fragrant, good quality honey.

Roses and other plants with double flowers tend to block access to their pollen and nectar by closing off the stigma to bees and other pollinators.

Other Uses:
Cosmetics, Skin Care, Etc.

Rose water is much valued for sensitive, chapped, dry, cracked, or inflamed skin. For aging skin, wrinkling, or ruptured veins, rose water is much esteemed as a corrective and preventative. It is also good for cleansing and can improve the complexion and dislodge blackheads. And if this were not enough, rose water leaves you smelling like a rose, as well, for the heady aroma literally sticks to your skin.

Rose water is an ingredient in many commercial skin care and cosmetic products, such as cold creams and bath water additives. It is hard to imagine a more delightful and fragrant addition to massage oil than rose water, particularly considering its epidermic rewards. For super massage oil, one might combine rose water with the diluted (with an organic oil) essential oil of oregano. Rose water in spray bottles makes a delightful room freshener.

Essential Oil

Occasionally essential oil of rose is used in soaps, cosmetics, bath oils, and other toiletries. It is also found in skin care products and ointments, even pomades; however, due to the high price of the oil, it is commonly restricted to non-allergenic products.

Perfume

The essential oil is sometimes used in breath fresheners, but it is most often found in perfumery, where it is used as the base for fine perfumes. Rose perfume has long been associated with femininity and the attribute of pure love. A lovely and appropriate sentiment for its intoxicating fragrance.

Native Range:
The native range of the various species is noted with each individual rose.

Specific Rose Species:
The following roses are not native to the United States, although they have all been widely cultivated here, and many have naturalized, and in some cases, become a serious weed, especially in the eastern United States. In the drier western states, they are seldom a problem weed, so cultivation is less worrisome. All these species are easy to grow, and compared to hybrid tea roses, they are low maintenance, with the possible exception of pruning. Easterners should refrain, however, from growing the species identified in the text as invasive. Instead, they should wildcraft them since they have been widely naturalized in the eastern United States.

Roses native to the United States can probably be used in much the same way as the exotics listed here; however, except for some scant information of native tribal use, we have more knowledge and experience with these exotics. In addition, these exotics are generally much more fruitful and their hips are often larger than most native roses. Expect exceptions to this, however, and even small fruited natives can be delicious.

Rosa X Alba
(Maiden's Blush, Cottage Rose, York Rose, White Rose)

Maiden's Blush is the only hybrid rose profiled here. It is a cross between R. canina and R. damascena. It is fast growing, reaches 4 to 6 feet in height, and has erect stems. The flowers are ivory white to soft pinkish white. They may appear singly, as semi-doubles, or doubles. They are 2 to 3 inches in diameter and very fragrant. Blooming in mid-summer, the plant has gray-green foliage, with rather large leaflets and leaves. It is thorny, with hooked thorns of varying sizes.

Culture: Hardy to zone 4, it seems to be less invasive than many other roses.

Food: The hips (fruit) are large, juicy, and flavorful. Both the hips and the flower petals are eaten.

Essential Oil: Maiden's Blush is cultivated commercially for essential oil, particularly in southern France and Bulgaria.

Rosa Canina
(Dog Rose, Dog Brier)

Dog Rose is 3 to 10 feet tall and up to 10 feet wide, with arching stems. The flowers are whitish to rosy pink, with golden stamens, borne solitary or in small clusters. Each flower is up to 2 inches in diameter, sweetly fragrant, and blooms in early summer. The semi-evergreen foliage is medium green and aromatic. The leaflets are small, ½ to 1½ inches long, with prickly double serrated edges. Dog Rose is thorny, with hooked thorns. It has showy, scarlet to reddish purple hips, about ½ inch in diameter, with a smooth skin.

Culture: Dog Rose is hardy in zones 2 to 9, wants full sun to light shade, tolerates most soils, even wet soils, is particularly fond of rocky soils, and favors woodland edges. The variety *Inermis* is thornless.

Food: The hips have long been used for food. Archaeological evidence in Britain indicates their use dating back to at least 2,000 BCE. They are exceptionally high in vitamin C and give yields of 3 to 4 tons per acre.

These hips have known very wide popularity as a food and have been used in almost all the ways listed in the main text.

Hips are harvested in late summer when fully colored but still hard. They remain hard on the bush until softened by frosts, eventually becoming fleshy. The flower petals are also eaten and used for other purposes. The leaves are used for tea.

Medicine: The hips have been used in almost all the ways listed in the general discussion of roses. Today, the hip syrup is popular in some places as an immune system booster. A daily dose of one tablespoon of syrup should do it. It is grown commercially for nutritional supplements and for flavoring in pharmaceutical products. It might be prescribed as a mild astringent, a mild diuretic, or a mild laxative.

The name Dog Rose was born long ago from its use as a cure for rabies, which has proven completely ineffective.

Native Range: It is native to Europe, West Asia, and North Africa, and has been naturalized in the United States.

Rosa Centifolia
(Provence Rose, Cabbage Rose, Rose of One Hundred Petals), and the subspecies R.C. Muscosa (Moss Rose)

Provence Rose is typically 3 to 6 feet tall and can be as wide as 8 feet, and sometimes even wider. The plant itself has a loose, open form, and its flowers are double, pink, nodding, and very fragrant, each about 2½ inches in diameter. They bloom from June to July. The *Moss Rose's* flower is intensely fragrant, and there are a number of named varieties with mostly double flowers in whites, pinks and reds. Some varieties bloom repeatedly. The plant has large, broad leaves and leaflets, and is quite thorny. It is densely covered with green hairs that give it a mossy look.

Culture: These roses are hardy to zone 5, love sun, but tolerate light shade.

Food: The flower petals are a much-utilized flavoring for foods, beverages, and medicines. Even the flower buds are pickled whole and eaten.

Medicine: The Moss and Provence Rose are commercially cultivated for essential oil distillation in Europe, especially in southern France, and also in Morocco. The resulting oil is called "Attar of Roses." The oil is pale yellow and extremely fragrant, having over 300 constituents, primarily citronellol (18 to 22%), geraniol and nerol (10 to 15%), and phenyl ethanol (63%). The geraniol content renders the essential oil a strong antiseptic. This oil is commonly used in aromatherapy.

Perfume: Attar of Roses is used for fine perfumes due to its divinely sweet fragrance.

Native Range: They are native to the Caucasus, the region between the Caspian Sea and the Black sea.

Rosa Damascena
(Damask Rose)

Damask Rose is 5 to 8 feet tall, with spreading, arching stems. The loosely double flowers range from whitish to rosy red, but most are pink. They are 2½ to 3½ inches in diameter and are borne in very fragrant clusters that bloom from June to July. Leaflets are pale green and about 2½ inches long. The plant is very thorny and the thorns are hooked. It produces small red fruits in abundance, which are very showy and hang on through much of winter. There are Damask Rose cultivars, and it is a parent to various hybrids.

Culture: Damask Rose is hardy to zone 5, sun loving, but tolerant of light shade. It is not finicky about soil.

Food and Medicine: Both the Attar of Roses and the rose water from the Damask Rose are used as a flavoring as well as a potent antiseptic. Although the hips are small, they are fruitful, tasty, and employed for both food and medicine. The delicious flower petals are also utilized for food and medicine, the aromatic leaves as a spice, and the new shoots, while still reddish in color, are eaten fresh, steamed, or boiled.

Miscellaneous Uses: Damask Rose is commercially cultivated for its essential oil in southern Europe, particularly in the east, with Bulgaria the center of production. This region accounts for much of the rose water in the marketplace (as a byproduct of the oil distillation). The primary constituent of its delightfully redolent oil is citronello, 34 to 55%. It also contains 30 to 40% geraniol and nerol and 16 to 22% stearopten. The principal use of these oils is for fine perfumes.

Damask Rose essential oil is excellent for skin care products as a skin softener that leaves the skin smelling delightful. The hips are still bright red in December and are popular for Christmas decorations.

Native Range: Damask rose is native to Asia Minor.

Rosa Eglanteria, syn. R. Rubiginosa (Sweet Briar, Eglantine Rose)

Sweet Briar is 3 to 12 feet tall and nearly as wide. It tends to be dense and erect in habit. The flowers are light to bright pink (or a variety of colors depending on the cultivar), about 1½ to 2 inches in diameter, and fragrant. They begin blooming near the middle of June, are borne solitary or in clusters, and can continue flowering until the hips first start coloring. The dark green leaflets are ½ to 1 inch long. They are scented with the aroma of apple, most noticeable after a rain or the morning dew. Sweet Briars are extremely well armed, being dense with stout hooked thorns. The smooth skinned scarlet hips are ½ inch in diameter and highly ornamental.

Culture: These roses are hardy in zones 2 to 9, prefer sun to light shade, and are tolerant of many soils, including alkaline.

Food: The hips are used for food, particularly in Norway, where they are often part of jams or jellies. The flower petals are also eaten and reputed to be tasty. They are popular in the Middle East as sweet treats.

Medicine: In Iran the flowers are used to remedy colic and diarrhea.

Agricultural Uses: Sweet briar is useful as a living stock fence. Planted 3 to 4 feet apart, it makes a dense, impenetrable barrier.

Native Range: They are native to Europe and Western Asia, and have been naturalized in the eastern United States and in parts of the Pacific States, particularly in the coastal zones.

Rosa Gallica
(Apothecary Rose, French Rose)

Aporthecary Rose is 3 to 6 feet tall and half as wide, if its roots do not creep out further and sprout. The stems are slender. The flowers range from bright, rosy pink to scarlet, and are the source of numerous cultivars and hybrids in a variety of colors. They are very fragrant, 2 to 2½ inches in diameter, borne in singles or doubles, and bloom from May to July. It has attractive blue-green foliage, and its leaflets are smooth, thick, and grow up to 2 inches long. The stems are thorny, with both stiff and weak thorns. The hips are dull red and about ½ inch in diameter.

Culture: Apothecary Roses are hardy to zone 4, prefer sun to light shade, flowering best in full sun. They are tolerant of a wide range of soils and drought.

Miscellaneous Uses: The Apothecary Rose is commercially cultivated for the essential oil, primarily in Eastern Europe. The distilled oil is marketed as "Attar of Roses" or "Essence of the Balkans." It is used for flavoring in food and drinks, or medicinally in skin care products and toiletries, but its primarily use is in fine perfumes. The fruits are eaten and used medicinally, as are the delicately flavored flower petals. The flowers are popular in potpourri.

Native Range: They are native to Europe, Western Asia, and North Africa, and have naturalized in the eastern United States.

Rosa Moschata
(Musk Rose)

Musk Rose is a fast growing, climbing or rambling rose, that can reach 8 feet or more in height, with strongly arching stems. The flowers are ivory or creamy white, 1 to 2 inches in diameter, with a rich, musk fragrance. The leaflets are ½ to 3 inches long, with finely serrated edges. They have both straight and slightly curved thorns. The fruit is small.

Culture: The cold tolerance is not well known; however, it should extend to zone 7. These roses like sun to part shade, favoring late afternoon shading. They can endure drought and tolerate many soils, though they tend to favor rocky places.

Food: These roses are grown commercially for their distinctive essential oil. The flower petals are eaten for their delightful deliciousness, both fresh or lightly cooked. The new shoots are also dined on, and are consumed fresh, steamed, or boiled.

Native Range: These roses are native to southern Europe, western Asia, and North Africa, and have been naturalized in much of north and central Europe, as well as in the United States.

Rosa Multiflora
(Tea Rose, Japanese Rose, Giang Wei Hua)

Tea Rose is very fast growing and dense, 3 to 12 feet tall by 10 to 15 feet wide. A very large, rambling vine or shrub, it may climb to 20 feet or form an arching, sprawling, impenetrable mass. They produce beautiful white to pink flowers in a profusion of large, fragrant clusters, often in double clusters. They bloom for a long time, beginning in June. Leaflets are about 1¼ inches long, with serrated edges. Some forms are thornless or the thorns are merely slender prickles. The scarlet hips are small, about ¼ inch in diameter, but are produced in great abundance, hang on into winter, and are attractive. There are a variety of cultivars.

Culture: Tea Rose is hardy to zone 5. It likes sun to part shade, does well on any good soil, but prefers clay. It is fairly drought tolerant and does well on steep banks. Diseases are seldom serious, but it is susceptible to mildew and spider mites.

Food: The hips are used as food in a variety of ways, sometimes eaten fresh or added to jams, jellies, syrup, and wine. Both the young leaves and the baby flowers are parboiled and eaten. The pith of the spring shoots are also used for food.

Medicine: For many centuries, this rose has been esteemed in Chinese folk medicine as few other species have. The flowers, hips, seeds, leaves, and roots have each provided remedies of some sort. In the past and present, a decoction of the flowers is valued for its cooling influence on internal symptoms associated with high summer temperatures. It is used to calm a jittery stomach, cure diarrhea, allay dry mouth and excessive thirst, release a tight chest, and check spitting blood and vomiting. It was once prescribed for the treatment of malaria.

The seeds have been used as a diuretic and a mild laxative. The hips are used in Chinese folk medicine as a tonic or to relieve excessive gas. The astringency of the hips is reputed to be healing and pain relieving for sores, wounds, sprains, and various other injuries. The roots' astringency has been put to work on healing skin disease, wounds, and ulcers. It also serves to regulate the excessive flow of any bodily secretions, in particular urine. The leaves have been used to treat ulcers and swelling.

Agricultural Uses: Tea Roses provide habitat for some beneficial insects, particularly parasitic wasps. They are often planted around orchards that are attacked by leaf roller insects to minimize leaf roller populations; the wasps lay their eggs in the rolled leaves.

Other Uses: This rose has been used as a functional, structural element, acting as a windbreak component, a living stock fence, a barrier to deer, or a crash absorber along highways. It has also been widely planted for erosion control on steep banks and road cuts.

Ecological Functions: Despite its aggressive weediness, it is rated as a fairly good habitat and shelter for wildlife. Certainly various species of birds are fond of the hips and also use it as a nesting site. Due to its invasive nature it should never be planted within miles of natural systems, if ever.

Native Range: Tea Rose is native to Japan, Korea, and China, and has been naturalized in the eastern United States and Canada. It is very invasive, and like other escaped cultivated roses, the seeds are spread near and far by birds.

Rosa Rugosa
(Rugosa Rose, Wrinkled Rose, Mei Gui Hua)

Rugosa Rose is fast growing and typically dense and shrubby, reaching 2 to 8 feet in height and 4 to 6 feet in width. Although uncommon, they occasionally reach 15 feet tall and 20 feet wide. The flowers are dark pink with a golden central disk and stamens and are very fragrant. The named varieties range from pure white to a deep rose purple and may not all be as fragrant. The flowers are 2½ to 3½ inches in diameter and borne solitary or, occasionally, in clusters. They bloom from early June to August, and sometimes as late as September or October. The leaflets are a glossy and wrinkled dark green. They are thick and 1 to 2 inches long. In fall they become yellowish or deep rosy orange, although often only briefly. They have many straight thorns, and the hips, which develop over a long period from June to September, are brick red to bright orange.

Culture: Rugosa Rose is hardy to zone 2. It likes sun to light shade and prefers sandy, well-drained soil with high organic matter content. It can do well in both pure sand and heavy clay, and most any soil in between. They prefer neutral to slightly acid soil, but are pH-adaptable, although they may yellow if soil is too alkaline. They are saline tolerant and very drought tolerant, but thrive with regular water. Rugosas are easy to grow and are rarely bothered by pests or disease, although like all roses, they may be colonized by aphids. They are very wind resistant and are tolerant of pruning.

The variety *Frau Dagmar Hastrup* produces hips abundantly. However, varieties with double flowers yield less than the species, probably due to the density of the petals barring entry to pollinators.

Medicine: Rugosa Roses are cultivated commercially in the United States for natural nutritional supplements, with the hips serving as an extremely rich source of vitamin C. The large, fleshy hips, succulent flower petals, and aromatic leaves are used in food and medicine. Rugosa Rose is the only rose listed as an official medicine in the *Pharmacopoeia of the Peoples Republic of China*, meaning the treatments it recommends have been confirmed by science. Its primary use is to correct stomach disorders such as chronic gastritis, the resulting diarrhea, and gastric neurosis. It is also officially prescribed for female complaints, such as irregular periods, leucorrhea, and inflammation of the breasts due to cysts obstructing breast ducts. It is also used for rheumatic pain and dysentery.

In TCM, the Rugosa Rose is said to regulate the flow of energy through the body, principally by normalizing the circulatory system. It has long been applied as a treatment for diseases of the blood. Prescribed as a general treatment for liver disease, it has a soothing influence on liver pain, and is taken as a tonic for the liver and spleen. In animal studies, the essential oil from its flowers was shown to increase liver bile flow. It is reputed to calm a restless fetus, and has long been a remedy for sprains, swellings, and inflammations. The most common way it is ingested is as flower petal tea.

Other Uses: Rugosa Roses are often planted to control erosion. The density of their foliage makes them good for windbreaks. *Native Range:* Rugosas thrive from sea level to 9,000 feet (in the Colorado Rockies), with the exception of at least one of the named varieties, *Blanc Double de Coubert,* which has not done well above 8,000 feet. Rugosas are propagated by stem cuttings or root suckers. They are native to northern China, Korea, and Japan. They have been naturalized in the northeast United States and eastern Canada, spread mostly from seed by birds to sandy seaside beaches.

Rosa Spinosissima, syn. R. Pimpinelliflolia (Scotch Rose, Scotch Briar, Burnet Rose)

Scotch Rose grows rapidly, reaching 3 to 4 feet in height, and spreads widely with aggressive suckers, often forming mound-like thickets that are upright, dense, and symmetrical. The flowers bloom in late May or early June. They range in color from creamy white to light yellow, are sometimes pinkish, and are ¾ to 2 inches in diameter. Leaflets are very small and rounded with saw-toothed edges. They are very thorny with both large and small needle-like thorns. The hips are very dark purple or brown, nearly black, and are ½ to ¾ inch in diameter. They produce abundantly and ripen in September.

Culture: Scotch Rose is hardy to zone 4, and possibly colder. It prefers full sun to light shade. It is easily grown in any decent soil. Scotch Rose is common near the sea, on cliff tops, in dunes, and among rocks.

Food: The hips of Scotch Rose have been much esteemed for food because they are especially sweet and flavorful, and because the plant is highly fruitful. The leaves are used for tea.

Others Uses: The aggressive rootstock of the Scotch Rose makes a good soil binder, and it has been planted specifically for erosion control.

They also make a lovely violet dye that was once very popular.

Native Range: Scotch Rose is native to the British Isles (the only rose native to Ireland) and most of Europe as well as Asia. It is an aggressive naturalized weed in the northeast United States.

Rosa Villosa syn. R. Pomifera (Apple Rose)

Apple Rose is variable in size, ranging from 1½ to 8 feet tall. Its growth habit is upright and the stems are slender. In bud, the flowers are deep red, but upon opening are clear pink, 1 to 2 inches in diameter, produced solitary or in small clusters. The aromatic leaflets are ½ to 1½ inches long. They are not particularly well armed, but the thorns are slender and straight. The large crimson hips are up to 1 inch in diameter and, unlike the other species represented here, are covered with tiny bristles.

Culture: Apple Rose is hardy in zones 2 to 9. It prefers full sun, tolerates many soil types, and is drought tolerant. In its natural range, it favors arid to semi-arid climates.

Food: The hips have been much used as a food. They are eaten fresh or dried, and can be used for making beverages, jams, sauces, puddings, sweet treats, and wine. The aromatic leaves are used for tea.

Ornamental: Apple Rose has often been planted as a hedge.

Native Range: It is native from Central Europe to Iran.

SAMBUCUS SPECIES
(Elderberry)
[Caprifoliaceae Family]

Elderberries leave taxonomists in a state of confusion. Scientists have yet to find common ground on the identification of species for the blue-black fruited elders native to the United States. The recent trend seems to be clumping most native United States species under the name of the European elder (Sambucus nigra), which seems an illogical and somewhat disrespectful solution, delegating all the American species to merely a subspecies of Sambucus nigra.

Elders are all very much alike on the one hand, and yet they have numerous minor differences. This explains the difficulty in classifying the species. These differences are likely related to the widely varying habitats, climates, and soils where they are found. All elders of temperate climates have pinnate leaves with numerous long, narrow leaflets suggestive of the tropics. Their flowers are borne in clusters (cymes) that vary from flat-topped to somewhat dome-shaped. The flowers range from white to light pink or soft yellow and tend to have a honey-sweet fragrance of varying potency.

For centuries in Denmark, elders have been considered essential habitat for elves. Although the subject is controversial, these shy creatures have so far refrained from comment. In the meantime, if for no other reason, people are planting elders just to provide for elfin habitat.

Sambucus Nigra

For our purposes, red-fruited elders are not profiled here due to their higher levels of toxicity, less tasty fruit, and the fact that their use medicinally is much less significant. That said, they are certainly not without value.

Elderberry in General
Culture (Exceptions noted under individual species):

Elderberries prefer sun or partly shady locations, although the more sun the fruit is exposed to the sweeter it becomes and the greater the yield. Blackbead elder appears to be the most shade tolerant variety. Elderberries like rich, loamy, well-drained soils, but are fairly adaptable. In the northern part of their range they prefer slightly acidic soil, while in southern regions they prefer a neutral to slightly alkaline soil. In far western areas, blue elderberry and blackbead elder tend to be the most tolerant of acidity, while desert elder is no doubt the most tolerant of alkalinity. Desert elder is tolerant of shallow or heavy soils. All elderberry varieties give their best performance on persistently moist soils, but tend to tolerate some degree of drought. Blackbead elder is the most water loving variety, but on shady sites can tolerate some dry soil.

Fertilization is not required for elderberries. Some wild plants can be self-fertile, but few if any cultivars are sufficiently self-pollinating. When using varieties, two or more varieties of the same species should be grown for cross-pollination.

Each spring numerous new suckers pop from the ground and grow up to 7 feet tall in their first year. During their second year, they start producing fruit and may continue to do so for three or four years; however, it is recommended that these shoots be cut to the ground at the end of their third year once they have become dormant in the winter. By the fourth year, they will be producing poorly if at all.

At the end of each year, while they are dormant, pinch the suckers back a foot to promote branching and thus increasing fruit yields. To keep track of a sucker's age, date each one with a bit of paint.

Elderberries transplant readily and are easy to grow. They are rarely troubled by pests or disease. The closest thing to a real pest is the elderberry aphid, although ravenous fruit-eating birds should be added to the list. Netting may be required if you hope to have some fruit for yourself. Elderberry borers may attack but seldom do any significant harm. At times, they may be troubled by cankers, leaf spots, or powdery mildew.

Propagation by either hardwood or softwood cuttings root well and is the most common way they are started. They can also be propagated by seeds or suckers (root division), although this is less common.

In most cases, flower clusters bloom irregularly, with those on the outer edges of the plant blooming first. Flower clusters are somewhat fragile so handle them carefully. Do not harvest the flowers unless they are dry. Any moisture from dew or rain will turn them black as they dry. The flowers can be shade-dried, but good air circulation is essential. If dried in an oven, a temperature of 100° to 200° F is recommended. Once dried, the flowers are separated from their stems by hand or machine friction. This should not be necessary if the bag method of petal harvesting (see below) is used. It has been reported that harvesting the flowers promotes increased flowering the following year.

Harvesting the flower clusters will result in the plant yielding no fruit. However, it is quite possible to get both flowers and fruit. Once the flowers have been pollinated, paper bags can be fitted over each cluster. When the flowers have matured, shake the bags so the petals are released from the cluster, thus giving you the flowers while leaving the fertilized parts behind to produce fruit. Once dried or cooked the unpleasant smell and taste of the fruits vanish.

Even when ripe, elderberry fruits remain firm. They are ready for harvest once they have fully colored. Harvest the entire cluster, and if you are planning to dry the fruit, you can hang the clusters from a clothesline until thoroughly dry. Some people prefer to strip them from the stalks and place them on flat trays in the sun to dry.

When harvesting the fruit for cooking, rather than drying, keep them in a cool spot until they can be refrigerated or frozen or cooked immediately. Before doing so, strip the fruit from their stalks using an implement like a table fork to separate them. Once refrigerated, it is best to cook them within two or three days. Otherwise they should be frozen.

Food:
Fruit
Elderberry fruits are exceptionally rich in vitamin C, offering 43% of the daily requirement per ½ cup. They are also a very good source of B vitamins and vitamin A as well as calcium, potassium, and iron. They are unusually high in protein content for a fruit and contain numerous bioflavonoids. In addition, elderberries are an awesome source of fiber, offering 5 grams per ½ cup.

Elderberry fruits have been used in a variety of ways in foods and drinks, either dried or cooked, and typically sweetened to taste. Although they require pectin for jelling, elderberries have been commonly used in jellies, jams and preserves, sauces, chutneys, soups, syrup, ketchup, pies, puddings, relishes, stews, sorbets, and food coloring, as well as in pancakes and muffins. Elderberries are also used in juices, wines, and vinegars.

The fruits are quick to ferment and produce a natural sparkling wine. Once bottled, there is a risk that gas can build up and cause the bottles to explode if the fermentation process is discontinued too early. To avoid excessive foam when the bottle is opened, the wine should always be served cold. The fresh fruit is not particularly tasty and has an unpleasant odor. Once dried or cooked, the taste improves significantly and the smell goes away. The dried fruits keep well and can be reconstituted in boiling water. Sun-drying provides the best flavor.

Flowers

The flowers add their own special touch in the kitchen, and they are a rich source of vitamin A, potassium, copper, pectin, and flavonoids. After removing the bitter stems, the fragrant flower clusters make delicious fritters when dipped in a sweetened batter and fried in oil. The flowers also add flavor, lightness, and nutrients to cakes, muffins, and pancakes. They are used in pies, to make a cider-like wine, and to make a vinegar that is used as a condiment. They are also used to flavor other types of wines, cakes, syrups, beverages, and ice cream. They make a delightful aromatic tea, and the buds and new shoots can be pickled. When used as an ingredient in salads, the fresh flower clusters are held over the salad and shaken to dislodge the flower parts. These clusters are never washed for fear of destroying their delicate texture and delicious ambrosial fragrance.

Medicine:

The bulk of scientific inquiry into the medicinal value of elderberries has been done on the old world species. Since the various North American species, in most cases, are so similar to Sambucus nigra, their medicinal properties may generally be considered interchangeable. The chemical constituents and pharmacological actions of elderberries are quite complex and are not yet fully understood. However, many of its folk medicinal applications have been shown to be effective through scientific research.

Being especially rich in flavonoids has much to do with elderberry's healing properties, including its very effective scavenging of free radicals and its suppression of oxygen ions that damage tissues and promote inflammation.

The flavonoid quercetin in elderberries can strengthen the immune system, and it protects LDL cholesterol from oxidation, thus reducing the chance of heart disease. Quercetin also obstructs the buildup of sorbitol, an enzyme that can damage the eyes, kidneys, and nervous systems of those that suffer from diabetes. It may also suppress cancer cells and tumor growth.

Flavonoids are water-soluble plant pigments that act as antioxidants. Flavonoids, quercetin in particular, along with elderberry's lignans, are strongly antiviral and help fight off the influenza virus. Research indicates that they may also prove effective against other viruses, including Epstein-Barr and other forms of herpes, type 1 poliovirus, and HIV. Flavonoids and lignans also work together to treat upper respiratory tract infections, including sinusitis and sore throats.

There are over 200 different types of lignans that provide numerous important functions in the body, including the protection and repair of liver cells. Elderberries have these biological liver-protection properties.

Bioflavonoids also protect vitamin C (an antioxidant) and increase its absorption rate and potency. Vitamin C elevates the body's production of glutathione (GHS), a highly potent cellular antioxidant that protects the DNA structure.

Glutathione offers a powerful cellular defense against carcinogens. It appears to slow the process of HIV advancing to AIDS, and it acts as a viral replication preventative. Glutathione is also involved in conveying amino acids across cell membranes. The trace element selenium (another antioxidant) must be sufficiently present in our diet to activate GHS. Foods richest in selenium include Brazil nuts, herring, salmon, whole grains, wheat germ, whole wheat bread, and brewer's yeast. However, anything over 0.15 milligrams of selenium can be toxic. Two hundred micrograms is recommended daily, although about 80% of Americans are believed to consume less than 100 micrograms.

Another bioflavonoid, rutin, is also involved in elderberry's healing properties. It acts as a free radical scavenger and is very important in the treatment of capillary fragility and varicose vein conditions. Rutin is antibacterial, anti-inflammatory, and antiviral. It is the synergistic effect of the elderberry flavonoids, quercetin and rutin, along with its lignans and vitamin C and the resulting increase of GHS, that accounts for much of the healing power possessed by elderberries.

Other medicinal contributors include the anthocyanins found in the fruit. They are a group of reddish-purple pigments and an important class of flavonoids. They stimulate the immune system and are a strong anti-inflammatory. They offer protection from diseases like cardiac artery disease as well as diabetic retinopathy, Alzheimer's, and arthritis.

Choline, another constituent of elderberry, is a precursor to the neurotransmitter acetylcholine that plays an important role in brain functions that effect memory, movement, coordination, and stamina. Choline also is an essential nutrient for protein, carbohydrate, and fat metabolism. It maintains the integrity of cellular membranes and expedites the movement of fats in and out of cells. It is valuable for calcium retention and for storing vitamin A in the body. It is a significant aid in repairing damage to the brain, heart, and kidneys, and lowers levels of the toxin homocysteine in the body.

Elderberries help dispose of cellular wastes, especially for jaundice. It offers pain relief for forms of rheumatism and arthritis, as well as headaches, including migraines. It lowers high blood pressure, promotes resistance to infections, and strengthens the immune system.

Fruit

Elderberries contain a dozen or more antibacterial and antiviral compounds, and nearly a dozen more that are antifungal. Additionally, the fruit can be applied as an analgesic, an anti-inflammatory, an antipyretic, a diaphoretic, a diuretic, and a laxative. The fruit also promotes production of the protein cytokine that is generated principally by white blood cells. Elderberries are valuable in the treatment of most upper respiratory tract infections. They have been prescribed effectively to treat influenza, sinusitis, hay fever, fever and chills, coughs including bronchitis, colds, sore throats, and tonsillitis. They can also help expel excess phlegm.

A flu and cold tonic can be made by cooking the fruit down to a syrup that soothes sore throats and helps open lung passages. A tincture can be made using 5 ounces of dried elderberries to a quart of 100 proof vodka. Add the dried fruit to a quart jar and then fill the jar to the top with vodka and tighten the lid firmly. The next day add vodka to the top again and let it sit in a cool, dark place for about a month and a half. Then strain thoroughly through many layers of cheesecloth.

Flowers

The flowers are rich in vitamin A, potassium, copper, pectin, and bioflavonoids, including rutin and quercetin. The medicinal actions of the flowers are anticatarrhal, analgesic, antibacterial, antipyretic, antirheumatic, antispasmodic, antiviral, and diaphoretic.

Leaves

The leaves contain bioflavonoids such as quercetin in many of its glycosidal forms as well as rutin. In addition, they contain fatty acids and tannic acid. Medicinally they are used as an antipyretic, a diaphoretic, a diuretic, an emollient, an expectorant, a purgative, and a vulnerary.

Elderberry leaves are primarily used topically as a poultice to treat burns, bruises, chilblains, scalds, sprains, swollen or stiff joints, tumors, and other wounds. Since the fruits and flowers perform as well or better than the leaves for internal applications, it is probably best to employ the leaves only for external use. This is suggested due to cyanogenic glycosides found in the leaves which are toxic when ingested.

Warning and Considerations:

While there are no reports of side effects, health hazards, or drug interactions with proper medicinal use of elderberry treatments, there are nevertheless concerns that should be noted. Elderberries contain cynogenic glycosides that release hydrocyanic acid—a poisonous solution of hydrogen cyanide gas (HCN). Many plants manufacture HCN as a deterrent to wildlife predation, including a few dozen common edible plants. The ripe fruit and flowers of elderberry plants are free of HCN; however, immature fruit, seeds, and leaves do contain it. The bark possesses an even greater volume.

Over millennia, as new varieties of cultivated crops were

produced, cyanogenic glycocides were increasingly reduced and even eliminated entirely through breeding. Where plants receive desirable amounts of moisture through the year, the presence of HCN can be negligible. As C. Leigh Broadhurst PhD and James A. Duke PhD point out in an article, when one's diet includes sufficient complete protein, HCN is largely disabled so foods that contain it can be safely eaten in moderation.

The fruit and flowers contain the alkaloid sambucine that can cause nausea. Cooking or drying eliminates this effect. Cooking is a good time to also strain out the seeds, although cooking itself may render them harmless. Elderberry stems make excellent flutes and pea shooters, but be forewarned that the toxic bark must be removed, especially around the mouthpieces. Children, in particular, need to understand this as a number of such poisonings have been reported involving children.

Agricultural Uses:
Insecticide
Traditional farmers long ago recognized that elderberries offer a variety of important uses around the farm. Although actual scientific inquiry is scant, empirical evidence tells of several applications utilized by traditional farmers. Elderberry chemistry seems in many cases to support these applications.

Elderberry leaves (and probably the bark) are insecticidal and contain tannic acid. Both European and North American farmers have used the crushed leaves piled up around crops to protect against flea beetles, carrot flies, other winged pests, and peach tree borers. Cut branches with leaves attached have been used for the same insecticidal action.

If the leaves are infused in warm water, the filtered fluid can be used as a spray and is believed to control blight and reduce caterpillar predation on roses and other flowering plants. This same infusion may be used to control cucumber beetles and various species of root maggots. The leaves are also regarded as an effective repellent to fleas and lice. In bygone days, some farmers brushed their hard and soft winter wheat with foliage-laden branches of elderberry as a preventative against the viral disease yellow leaf spot (Helminthosporium tritici-vulgaris).

Elderberry flowers, as well as those still in bud form, are infused in water for an insecticidal spray against aphids. It is believed that the nauseating odor of the leaves and bark are responsible for the repellant effect on flies, mosquitoes, midges, and other insect pests. Rub the exposed parts of your body with the bruised leaves, and it will keep these pests at bay for about half an hour. To maintain your immunity you must repeat the process, but do not expect any hugs!

Commercial Beverages
There is a beverage market in Europe for elderberry flowers. Tasty herb teas, tonics, wines, and liqueurs can be made from the flavorful elderberry flowers. As a result, commercial elderberry orchards are becoming increasingly popular. Elderberries make a lot of sense as a valuable component of polycultural orchards.

Trap Plant
Another way that elderberries are used is to lure pesky birds away from commercial fruit crops like grapes, raspberries, and blackberries. Because these fruit-robbing birds prefer elderberries over more popular fruits, they are commonly planted around orchard perimeters.

Livestock Forage
Both the fruit and foliage of elderberries are consumed by livestock. However, the elderberries need to be protected from foraging until they are near maturity, and even then livestock should be managed and not overstocked to keep the plants in good condition.

Elderberries are good companions to chicken coops. They can be planted on the west side to provide important summer shade and can be irrigated by precipitation harvested off the roof of the coop. The chickens get shade, a place to roost outdoors, and good nutrition, experiencing solid weight gain from eating the fruit. They can also feed themselves without always relying on the farmer for food. There is, however, a laxative effect from consuming too much fruit. It is messy, but valuable for composting.

Compost
Elderberries make good companions to composting. Elderberry leaves are reputed to accelerate the decomposition process of the compost pile, thus making the finished compost ready sooner. Biodynamic farmers and gardeners worldwide plant elderberries near their compost piles. They are excellent soil builders and produce high quality humus. If planted west of the compost bins, they provide late afternoon shade and help to reduce dehydration of the pile during the hottest time of the day. When making compost, chicken manure supplies the nitrogen while Elderberry leaves supply the carbon—the two necessary elements, along with some moisture and air, for successful composting.

Veterinary Ointment
A veterinary ointment to treat sores on animals is made by adding elderberry flowers to warm melted lard.

Bee Forage

Elderberry flowers yield an abundance of protein-rich pollen that bees avidly collect. However, the amount of flower nectar they produce is negligible.

Other Uses:

Wind Breaks

Elderberries can be used as a component in windbreaks and shelterbelts when mixed with other trees and shrubs. Properly designed, planted (in windbreaks 10 feet apart), and established, they can help protect crops in a variety of ways or protect livestock from the extremes of heat and cold. With proper placement, they can also act as living snow fences, capturing snow to avoid drifts where they are unwanted.

Cosmetics

Elderberry flower water (the fresh, fragrant flowers distilled in water) is often used in the cosmetic industry. It forms the base of ointments, eye and skin lotions, creams (often made with buttermilk), oils, and inhalants.

Elderberry essential oil is rich in fatty acids (66%), which makes it an excellent skin softener that is very good for the hair as well. Elderberry flower water is a mild stimulant and is moderately astringent. It is valued for its capacity to improve the complexion, clear away blemishes, and soften and deeply cleanse the skin. It is used as a steam facial and is reputed to remove freckles.

Elderberry flower lotions provide all the benefits listed above and are also used to soothe sun and wind burns and relieve itching and other skin irritations. The lotion is effective for a long period of time, so preservatives are unnecessary in commercial elderberry products. The sweet, heady fragrance of the flowers is used in nature-based perfumes as well.

Dye

From First Nation peoples to the ancient Romans, elderberry fruits have been valued for their use as a hold-fast dye. The Romans used it principally as a hair dye to turn graying hair a lustrous black. Tribal groups of the Americas used it as a hair dye, but also used elderberries to produce lavender and purple dyes that they used to color fabrics and basketry designs.

Some Turtle Island cultures extracted bright yellow and orange dyes from elderberry stems for use in basketry and probably for clothes and other articles as well. A green dye can be made from the leaves, and when combined with chromium as a mordant, a gray-green color can be produced.

Music

Some Californian tribes knew the elderberry as the "Tree of Music." In spring they would cut off new stems (1 to 3 years old) for flutes, then hollow them out using a heated hardwood stem to burn away the pith. The hollow stems were also used for whistles, blowguns, and an instrument similar to panpipes. As many a country boy will remember, the hollowed stems also make pop guns and pea shooters. Often these same boys would become quite ill if they neglected to remove the bark around the mouthpiece.

Wood

Wood from the elderberry dries hard and has a fine, dense grain that cuts easily and takes a good polish. It has been used to make numerous items including mathematical instruments, mill cogs, butchers' skewers, spiles for tapping sap, toys, combs, and a variety of other small articles. Whittlers enjoy working with it green because it is very easy to carve.

Elderberries coppice vigorously after being cut or burned to the ground. Once they have re-sprouted, the leaf nodes grow further apart and each stem is straight, making them more valuable as a basic material to work with. Tribal groups living in what is today California used the coppiced stems for basketry, arrows, traps, and weirs. Coppice shoots also made the best flutes as well as clappers (a tribal percussion instrument).

Up to ninety percent of an elderberry's stem interior is pith. It is buoyant and very soft. In fact, it is the softest and lightest natural solid known. The specific gravity of elderberry pith is 0.09, compared to cork which is 0.24. Elderberry pith is used in microscopy as well as by scientific instrument manufacturers and watches makers as an absorbent cleaning material to sponge up oil and dirt from delicate mechanisms. The pith is also used in biological collections, serving as a gripping material for small specimens like flowers, leaves, or insects. Leave it to the ingenuity of children to discover another practical use for the pith. They found they could cut out the middle man when it came to fishing floats. A few green elderberry twigs work perfectly as a float.

Native Range:

Depending on species, elderberries are native in North American from coast to coast and from southern Canada to northern Mexico. The Sambucus nigra is native throughout Europe, though not on the islands.

Specific Elderberry Species:

As we look at profiles of the native elderberries of the United States, the degree of confusion involved in classifying them will become more apparent. Some authors do not even give a species name but simply lump them all together, no doubt to the consternation of botanists and taxonomists alike. Perhaps one day they will all be listed as varieties of a single common species, but until then, the following classifications are useful, primarily for research.

Sambucus Caerulea
[Syn. S. Nigra ssp. Caerulea, S. Mexicana, S. Glauca, S. Caerulea var. Velutina, S. Californica, S. Caerulea var. Neomexicana, S. Fimbriata]
(Blue Elderberry, Blueberry Elder, Capulin, Sauco)

Blue elderberry is a rapidly growing deciduous shrub or tree, often 6 to 15 feet tall, but capable of reaching 30 to 50 feet. Trunk diameters of 2 feet have been found on large, old trees. It can sucker profusely and will coppice readily. It is often almost as wide spreading as it is tall and produces white to creamy-white flowers in flat-top clusters (cymes) 2 to 10 inches in diameter. The blooming period is highly variable due to its huge native range, which stretches from Canada to Mexico and from sea level to 11,000 feet.

Culture: Blue elderberries do best with full sun exposure, but can also do well in partial shade. They prefer moist, well-drained soils, but tolerate moderate drought. They are at their best in sand or clay loams with good drainage.

Food: The fruits are produced in large, often prolific clusters. They are blue to purplish-black with a glaucous bloom and are ¼ to ⅜ inch in diameter. These sweet, juicy fruits are probably the best tasting of our native elderberries, with the possible exception of named fruiting varieties.

Ecological Functions: Sheep and cattle may browse them, but they are thought to be better for livestock after the first frost. Blue elderberries are reputed to be excellent for wildlife like deer and squirrels, and at least twelve species of birds are fond of the fruit, including pheasants, quails, and robins.

Native Range and Habitat: Blue elderberry is found in the southern portion of Vancouver Island and on the British Columbia mainland east to Alberta, and as far south as Southern California and New Mexico and in the mountains of Mexico's Sonoran and Chihuahua deserts. It grows in the Pacific Coast Range from north to south, on both sides of the Cascade Range, and through both the Sierra-Nevada Range and the Rocky Mountains.

Blue elderberries are common in riparian zones, and may also be found in moist meadows and canyons, growing in mixed conifer forests or in Aspen and Ponderosa pine belts. They are often associated with saskatoons, chokecherries, wheatgrass, and Bromes. Blue elderberry can sometimes be found in the piñon/juniper belts of the Southwest interior.

Sambucus Canadensis
[Syn. S. Nigra Ssp. Canadensis, S. Mexicana, S. Caerulea var. Mexicana]
(Sweet Elder, Common Elder, Pie Elder, American Elderberry, Canadian Elderberry)

Sweet elder is a rapidly growing deciduous shrub or small tree, typically 5 to 12 feet tall by 8 to 12 feet wide. On an ideal site, it can reach a height of 30 feet and will sucker readily into multiple stems, making it a good coppice plant. The flowers are a creamy white to a creamy yellow and are borne in clusters of up to 10 inches in diameter. The blooming period is variable due to its very large native range, but in many areas, blooming begins in April or May and continues on into July or August. The fruit ripens in clusters of blue, deep purple, or black berries that are coated with a white, waxy film. Each fruit is about a ¼ inch in diameter. A number of cultivars have been developed for improved fruit quality and higher yields. These are grown commercially, mainly in the dry south Canadian interior or in the northern interior of the United States, where few fruits can be cultivated successfully for commercial production.

Culture: Sweet elders are most productive in full sun, but are well adapted to partial shade. They are generally found in moist, rich, slightly acidic soil, though they seem to tolerate some drought. They are hardy to zone 4 and milder parts of zone 3, ranging from sea level up to 4,000 feet or more in the south. The varieties *Adams, Nova,* and *York* have thrived when planted in the White Mountains of east-central Arizona at an altitude of 8,400 feet.

Varieties:

Some of the best varieties of sweet elder are listed below. All of them need a pollinator, so plant two or more varieties.

Adams #1: Adams #1 produces remarkably large fruit and fruit clusters. It is very productive with vigorous growth.

Adams #2: Adams #2 produces large fruit, but not as large as Adams #1. However, its fruit clusters are larger. It is a bit more productive than Adams #1. It is pest and disease resistant and grows to 6 to 10 feet tall.

Johns: Johns produces large fruit and fruit clusters. It is productive and disease and pest resistant. It is very soil adaptable and grows vigorously up to 10 feet.

Nova: Nova produces large fruit that ripens evenly, making harvest sorting less time consuming. It is productive, but can sucker vigorously. It grows from 6 to 8 feet tall and is an open pollinated seedling of Adams #2.

York: York produces the largest fruit of any of the other cultivars and in large clusters. It is very productive and probably the tallest growing and widest spreading cultivar.

Ecological Functions: Livestock are often fond of browsing sweet elder, but in some areas, cattle will have nothing to do with it. At least forty-five species of birds are attracted to the fruits, including wild turkeys, prairie chickens, quails, pheasants, doves, and grouse. Several other wildlife species consume the foliage including moose, elk, and deer. Other elderberry fruit eaters are rabbits, woodchucks, squirrels, and chipmunks, that also eat the bark.

Natural Range and Habitat: Sweet elders' natural range is from Manitoba east to Nova Scotia in the north, and south to Texas in the west and Florida in the east. They grow in riparian corridors, margins of forest and woodlands, bottomlands, swales, and low depressions.

Sambucus Melanocarpa
[Syn. S. Microbotrys var. Melanocarpa, S. Racemosa var. Melanocarpa]
(Blackbead Elder, Black Elderberry, Mountain Elder, Flor de Sauco)

Blackbead elder is a rapidly growing deciduous shrub that typically reaches 3 to 12 feet in height, but can attain heights of 30 feet. It tends to sucker and form colonies. It has white flowers formed in 2 to 3 inch diameter clusters that bloom in May or June through July or August. It has black to dark purplish fruit and lacks a glaucous coating. Each fruit is about ¼ inch in diameter or smaller.

Culture: Blackbead elder grows in sun to part shade and can thrive in fairly deep shade, though the fruits are compromised. It generally requires moist, even wet soil. However, it can do well at times in dry shade.

Ecological Functions: Blackbead elder is generally avoided by livestock, although it is palatable to sheep in some areas. Livestock find it more desirable after the first frost. Birds and other wildlife on the other hand readily utilize it.

Native Range and Habitat: Blackbead elder is a mountain plant. Its native region runs from British Columbia to Alberta in the north to California and across to New Mexico in the south. In the Oregon Cascades, blackbead elder starts at 2,500 feet on the east side and goes up in altitude from there. In the Rockies, it grows from the outer foothills to the edge of the alpine zone. In the Sierra Nevada Mountains it grows at altitudes between 6,000 and 12,000 feet. In Arizona's mountains it is found above 7,500 feet. It is often found in riparian areas, canyons, meadow edges, and open conifer forests.

Sambucus Mexicana
[Syn. S. Coerulea var. Arizonica S. Coerulea var. Mexicana, S. Canadensis var. Mexicana, S. Velutina, S. Coruacea, S. Orbiculata]
(Desert Elder, Mexican Elder, Blue Elder, Tapiro, Sauco, Capulin Silvestre)

Desert elder is a semi-evergreen shrub that goes deciduous during the hottest, driest part of the summer. It grows 6 to 15 feet tall, but can reach a height of 35 feet with a width of 15 to 20 feet. It is moderately fast growing, but grows rapidly with ample irrigation. It has white flower clusters, though sometimes they can be cream or pale yellow, 4 to 8 inches in diameter. They generally bloom in March or April to May or June, though sometimes as late as August. They tend to bloom after rain, and when circumstances are just right, they may bloom over a long period. Desert elder fruits are purplish black to blue or occasionally white and up to ¼ inch in diameter. Typically birds avoid the white fruit.

Culture: Desert elder prefers sun to part shade. It is probably at its best on moist, well-drained loams, but is known to tolerate coarse sands or gravels and even poorly drained heavy clays. It tolerates some drought (especially in summer dormancy), probably more than other species, and is the elder best adapted to high temperatures. For best results, it may need to be irrigated in the dry season when soil moisture is scant. On the other hand, it does well in soggy soils.

Ecological Functions: It is used for food by wildlife including over a dozen species of birds.

Native Range and Habitat: Desert elder is native from central California to west Texas and south into Mexico. In Arizona it resides at altitudes between 1,000 and 4,500 feet. In central California it is seen at low to mid altitudes. It is a component of a variety of plant communities and is often found in the creosote belt and occasionally among piñon/junipers. It is common in the lowlands along riparian corridors and through arid grasslands as well as swales, arroyos, and valleys.

Sambucus Nigra
(European Black Elderberry)

European black elderberry is a fast growing deciduous shrub or small tree that grows from 6 to 35 feet tall. Its flowers are a creamy white or yellowish and grow in clusters 5 to 8 inches in diameter and tend to bloom in June and July. They are more fragrant than our American elderberry flowers and have a sweet muscatel aroma. In addition, they remain fragrant even when they have been dried. The shiny black to purplish black fruits (rarely white or pink) are borne in large clusters and each fruit is ¼ to ½ inch in diameter.

Culture: European black elderberry grows well in sun or part shade and prefers moist, rich, loamy soil, but is largely adaptable to different soils as long as moisture is adequate. Like most elderberries, it is easy to grow, although its performance in the United States can be disappointing compared to native varieties. Still, European black elderberry can naturalize in the United States when conditions are favorable. Perhaps its greatest challenge in the United States is its high susceptibility to spider mite infestation in hot weather.

Food: Both the flowers and the fruit are used in Europe to make wine, and the leaves are used for herbal tea.

Miscellaneous: Europeans have found many uses for black elderberry's leaves, flowers, fruit, and bark in inhalers, lotions, dyes, and insecticides. The wood is used to make a variety of small articles.

Varieties: European varieties of Sambucus nigra are now becoming available in the United States. All of them need a pollinator. A few are listed below:

Alleso: This is a commercial variety with productive fruit yields, growing to about 10 feet tall.

Haschberg: Haschberg produces large fruits borne in large clusters, is very productive and fast growing, reaching about 10 feet tall.

Korsor: This is a commercial variety valued in the marketplace for its medicinal properties. It is a good pollinator that grows to about 8 feet tall.

Samdal: Samdal bears large fruit clusters and tends to sucker annually.

Sampo: This is another commercial variety with productive yields, growing up to 10 feet tall.

Native Range and Habitat: European black elderberry is native all over Europe except for the islands, and it ranges east to western Asia and south to North Africa. It often grows in woodlands and bottomlands, and even on cliffs facing the sea, lashed by salt spray. European black elderberry is found in a variety of plant communities.

Sambucus Simpsonii
(Florida Elder, American Gulf Elder)

Florida elders are typically large shrubs. This elder closely resembles Sambucus candadensis, and some believe that is actually what it is. (It is also possible it is a hybrid.) The black fruits are edible.

Native Range: It is native throughout Florida and west along the coast to Lousiana.

TRACHYCARPUS FORTUNEI
(Hemp Palm, Windmill Palm, Chusan)
[Arecaceae Family]

Hemp palm is an evergreen palm tree with a moderate to fast growth rate, reaching 20 to 40 feet in height with 10-foot diameter crowns. In arid or cold climates it grows more slowly and may get no bigger than 15 feet tall with 7-foot wide canopies. The fan-shaped leaves are nearly round in the canopy's outline and resemble a windmill, hence one of its common names. The dark, dull green leaves are 2 to 4 feet long and radiate out from the top of the trunk. The creamy white flowers are commonly unisexual, borne in uniquely attractive dust mop-like clusters. The fruit is round or kidney-shaped, blue to bluish, a ½ inch in diameter, and borne in clusters. The trunk is slender, but sturdy and erect. Its shape can make it appear swollen in the upper section because it is covered with a protruding woody base of old leaf stems and long, dark, shaggy hair-like fibers. The bottom portion of the trunk, however, is bare.

Culture:

The hemp palm prefers full sun to part or filtered shade and likes good garden soil, but is tolerant of alkalinity. It responds to annual applications of compost and/or organic fertilizer. It prefers moderate to ample watering but is generally drought tolerant once established. It is one of the most cold-tolerant of all palms, hardy to 0° F without damage. At minus 9° F the leaves may die back, but will recover quickly. Young palms need protection in cold winter zones. They can sunburn

Trachycarpus Fortunei

in the desert or in very hot, reflected sun. Once established, hemp palm endures heat, drought, wind, ephemeral wet feet, and neglect. It is a very self-sufficient palm, hardy to zone 7 and borderline in zone 6.

Seeds should germinate readily if in good condition. Most palm seeds have a short lifespan and ideally should be planted within several days of harvesting them. Seeds are best planted under glass in a well-drained mixture that might include peat, sand, and perlite. When a seedling's first leaf is mature, it should be stepped up, always taking care not to harm the tender roots. A seedling should be kept in 60% shade until it is ready to be moved to its permanent location.

Food:

In China, the young inflorescence still in bud stage are eaten and prepared in much the same ways as bamboo shoots.

Medicine:

In TCM, according to the *Pen Ts'ao* (Materia Medica 1578), the flowers, buds, and seeds are used to treat hemorrhage, as well as flux (excessive flow or discharge from an organ or cavity of the body). More than 400 years later it is still being used in TCM to stem these conditions.

Other Uses:
Fiber

Fibers from the trunk, leaf bases, and leaf sheaths are used to make a variety of articles. For example, its leaf-based fibers known as Chinese coir are used to make decorative hemp-like rope and other cordage, as well as brooms, brushes, and capes. The leaves are used to make hats, cloth, mats, and fans.

Wax

A wax can be extracted from the hemp palm fruit.

Native Range:

Hemp palms are indigenous to east and central China and northern Myanmar (Burma).

TYPHA SPECIES
(Cattail)
[Typhaceae Family]

In the continental United States three common species of cattail overlap. The hardiest and most prevalent United States species is Typha latifolia, followed by Typha angustifolia, and then Typha domingensis. In many temperate wetlands all three may grow near each other, and to the botanist's confusion, they readily hybridize with one another. Their economic usage is also essentially interchangeable.

The number of common names that have been used to describe these three species is astounding, no doubt because all are common natives of the northern hemisphere—from south of the equator almost to the arctic.

In the United States they have been called cattail, cattail flag, and reed-mace. Typha latifolia has also been called bulrush, cossack asparagus, common cattail, and broadleaf cattail. The most common name for Typha angustifolia is narrowleaf cattail. Both species might answer to soft flag or less often, nail rod. Typha domingensis has been referred to as southern cattail or tule cattail. To Hispanic Americans of the southwestern United States, cattails were usually called aguapá. Indigenous names include tabu'oo (Paiute), toiba (Washoe), apuk'we (Chippewa), k'ut (Cahuilla), olel (in coastal Salish), and pu (Mayan). To East Indians a similar species of cattail is called hagla or pun, and to the English colonists

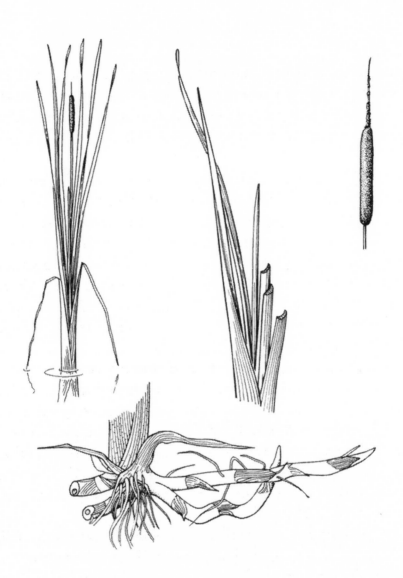

Typha Latifolia

elephant grass, possibly a reference indicating elephants are fond of it. To the Chinese it is pu huang, meaning golden rush.

Culture:

Cattails form extensive colonies in wetlands due to their aggressive, creeping rhizomes. They all have tall flat leaves that tend to spiral up their length in the wind, making them remarkably resistant to wind damage. The leaves of Typha latifolia are 3 to 6 feet long and up to 1 inch wide. Typha angustifolia's leaves are only about ½ an inch wide. The cattail flower's vertical stalk ranges from 3 to 9 feet tall. The downy velvet, cylindrical flower clusters, composed of thousands of flowers, appear at the top of the stalk. Uppermost are the male flowers in a short narrow cluster. Just below them are the females in a longer, fatter, rich brown cluster. These flowers are slightly separated from each other in Typha angustifolia, and both male and female clusters are a little smaller than those of the Typha latifolia.

When the male flowers mature, they are transformed into bright yellow pollen. The pollen is released in the summer for fertilization. After the pollen is liberated, a small portion of the naked stalk protrudes from the top of the female flower cluster. Once a female flower is fertilized, it begins to develop seed. Cattails bloom from May to July and, because individuals bloom in their own time, the flowering season may last for several weeks.

The seeds reach maturity in late summer and into fall. Gradually, over a long period, most will be released to the wind literally on gossamer wings due to the fine silky fiber attached to each seed. Each cluster contains thousands of tiny seeds which, as they are released, seem to mimic small clouds, drifting through the air.

Cattails are rhizomatous. The rhizomes lay just under the soil surface. They grow as horizontal stems in all directions and develop swollen nodes at intervals along the stems. Stalks shoot up vertically providing foliage and forming terminal flowers.

Cattails should receive full sun. Although semi-aquatic, they are also terrestrial in wet, muddy soils, preferring swampy marshes, ponds, or slow moving watercourse banks. They are generally propagated by division of the rhizomes. They can be planted by seed, although it is not as common. Seeds are sown in pots that sit in water. Cattails are easy to grow and almost impossible to eradicate as long as sufficient moisture is available. Be careful where you site them; they are very invasive. They can quickly take over a small shallow pond or an irrigation ditch. You might say with cattails that's an ecological mandate.

Food:
Cattails have been used as food from some dim mysterious prehistoric past to the present day. The young shoots are a good source of vitamin C. The rhizomes are rich in starch and sugar. When they are ground for baking, the flour is about 57% carbohydrate. In fact cattail flour is comparable to wheat, corn, and rice in carbohydrates and protein, with less fat. When cattail rhizome flour was fed to mice in the lab, they all put on solid weight with no ill effects.

Rihizomes
Research has demonstrated that cattails can reliably yield approximately one hundred and forty tons of rhizomes per acre, nearly ten times more than potatoes. The one hundred and forty tons in turn yields thirty-two tons of dried flour, far greater than what wheat, oats, rye, or millet can produce.

Even with aggressive rhizome harvesting, cattails often recover and fill in rapidly. With thoughtful planning, harvesting rhizomes can generate food while also improving nesting habitat for water fowl and increasing open water for fish.

Cattails were an important food plant to traditional tribal cultures and later to European settlers. Tribes across the United States, including the Iroquois, Abenaki, Yumans, Hopi, Acoma, Laguna, Apache, Paiutes, Pima, Cahuilla, Snohomish, Snuqualmi, Nisqually, and the people of San Felipe pueblo, found cattails to be a sumptuous food. Captain Lewis of the Lewis and Clark Expedition noted in his diary that the tribes of northeast Oregon enjoyed cattail rhizomes as a source of food.

In addition to the food values already mentioned, cattails are also very rich in valuable minerals, which they absorb from the water and mucky soil in which they grow, and can supply an abundance of wholesome foods all year long. Sprouting juvenile rhizomes, young shoots, immature flower clusters, male pollen, seeds, and mature rhizomes are all eaten at some time through the four seasons.

Rhizome Sprouts

A delectable cattail treat can be harvested in late winter or very early spring when the young rhizome buds begin to sprout. The ideal time to collect them is when the sprout has begun to swell but has not as yet pushed through the soil. Break or cut the young sprout from the mature parent, cut off the base and the green sprout tip, and peel it down to the tender core where the starch is concentrated. In her book *American Indian Food and Lore*, Carolyn Niethammer suggests keeping only the three or four innermost rings.

Dice them up and roast with meat, or boil the cores for

about ten minutes and serve them with coconut or olive oil for a delicious vegetable. The cores are also good for stir frying. Once they have been par-boiled they can be sautéed. Some people equate the rhizome sprouts' flavor to tapioca.

The cores can also be pickled to give them a long shelf life. A good tasting pickle can be made from the peeled, diced, and parboiled cores by adding them to hot vinegar, sealing them in a jar, and then giving them a few weeks to pickle.

Spring Shoots

Many New World tribes ate the young shoots of the mature rhizomes from the time they first appeared until they reached a height of 2 or 3 feet. The shoots can be harvested from the beginning of spring through the rest of the season. Just grab hold of the shoot near the base and pull it off the rhizome. They come free easily. Remove any leaf that may have come with the young shoot and peel it down to the succulent white core. This core will be about ½ an inch in diameter and about half the length of the shoot. They are good eaten as is or diced and added to salads.

The flavor is variable, ranging from mild but good to excellent and distinctive, and is very similar to cucumber. If placed in a covered bowl of water and refrigerated, the raw cores will stay fresh for a couple of days. Some people prefer them cooked. Bring a pan of water to a boil, add the diced cores, and simmer for about ten minutes. Serve as a vegetable with a dressing of olive oil and vinegar or coconut oil and tamari or add the cooked cores to soups or stews.

Russians and Europeans have considered cattail shoot cores a delicacy for a very long time. In the Southwest the Yuman of the lower Colorado River ate them raw or added them to cooked tepary beans and mashed them together.

Immature Flower Clusters

The flower clusters begin to form in late spring and continue forming until early summer. They are ready just before they start to break out of the sheath, but before the pollen begins to show. Cut them while still green and wrapped in their papery sheath of leaves. Shuck off the sheaths and cook immediately by boiling them just a few minutes. The outside of the de-sheathed cluster is then eaten with your hands. The soft flower buds are nibbled off, and the tough inner stem discarded afterward. The flavor is quite good and has been compared to olives and French artichokes. Various western tribes ate them raw as well as boiled. They also steamed them or added them to stews.

Some wild food enthusiasts scrape the young flowers off the stems and use them in casseroles. They can be frozen for long storage; shuck, boil five minutes, remove the flower buds from the stem, and freeze.

Pollen

Another source of tasty food becomes available in early summer when the cattail's male flowers mature into golden pollen. The brilliant pollen is easy to harvest by stripping it off into a paper bag. It is a rich source of protein, phosphorous, and sulfur. Euell Gibbons suggested its dazzling gold color indicates the possible presence of beta-carotene, probably in appreciable amounts. The East Indians make bread from the pollen.

Some First Nation tribes ate the pollen raw or with a little water as a gruel. Sometimes it was boiled then eaten or used to flavor other dishes. Otherwise, preparing the pollen as the tribal peoples did can be quite an elaborate process.

Today we need only a fine mesh screen or a strainer to sift the pollen. This removes the impurities, leaving a powdery material that looks like golden pastry flour. The pollen flour should be dried thoroughly before use. Today it is common to mix cattail pollen flour fifty-fifty with wheat flour for pancakes, muffins, fritters, and the like. Pure cattail pollen flour can be added to soups and stews to give them body, or as some tribes did, simply to make a sweet golden mush. Once the pollen has been sifted and dried, it can be stored in a closed container for future use.

Seeds

Cattail seeds begin to ripen in late summer and continue through the fall. Some indigenous tribes would liberate these tiny seeds from the cattail down and eat them. A large flat area was needed, such as a rock, upon which to pound the seeds and remove the down. Once the down is loosened, the fluff is spread more or less evenly across the rock surface, then set on fire. The fluffy down burns quickly then turns to ash. Thousands of seeds will be found in the ash of a single cluster. They are swept up and the ash is winnowed away. The seeds can be eaten raw or used to garnish other dishes or baked goods.

Rhizomes

Cattail rhizomes can be harvested for food anytime of the year; however, by late fall they are richer in starch which they maintain through the winter. They are at their peak nutritionally and flavor-wise during this period. They may be inaccessible in cold climates during winter if the water is frozen. Otherwise, the rhizomes offer an extraordinarily immense harvest potential, if you have the time. In fact, you should be able to collect enough rhizomes from a small area to feed a family through much of the year.

To harvest rhizomes rubber boots are a good investment.

Be forewarned, we are talking stoop labor here and processing the rhizomes can be somewhat tedious. On the other hand, the abundance of nutritious food they provide and their fine flavor will make the effort worthwhile.

Trace the stalk down into the water until you feel the rhizome, then slip your fingers under it and pull up gently until it is loose but still connected to the horizontal parts. Now loosen those parts and try to lift the rhizome whole out of the mud. Swish it around in the water to clean the bulk of the mud off and toss it in a container with some water in it. Then go for the next one.

As soon as possible, peel the rhizomes down to the core and compost the bark and spongy layer that is removed. Do not wait too long to do this because peeling becomes increasingly difficult as the roots dry out. Wash the cores thoroughly before preparing them to eat.

They can be chopped up and put into a large pot of cold water or grated directly into the pot. Some First Nation tribes crushed them between two stones before adding them to the water. If you have a heavy duty potato masher, you can crush them up in the pot of water. Since the purpose of the rhizome preparation is to separate and remove the unpleasant fibers from the starch, it is helpful to combine some of these techniques for best results. You should be diligent in this pursuit if you desire a grit-free product.

Once crushed, squash them further with your hands until the pulp is completely broken down, then let the starch settle to the bottom of the pot. Wait about half an hour and skim out the fiber. Pour off the water and repeat the process a few more times. This mess can be sifted through a clean cloth to help separate the starch from the fiber. What remains in the cloth is processed again in fresh water, repeatedly if need be, until the starch is fiber free. The result of this effort is a refined white material ready to use as flour. Use fresh or dry it and grind it into powder and store in a sealed container. The flour can be used to bake bread, muffins, biscuits, cookies, or just about anything calling for flour. Use the flour by itself or mix 50/50 with any kind of flour your recipe calls for. A low-tech fiber sifting apparatus should not be hard to invent or construct.

One of Carolyn Niethammer's recipes in her book *American Indian Food and Lore* sounds especially delicious, and it bypasses all the refining needed for flour. Once the bark and spongy parts have been cut away from the rhizome's starchy core, it is washed and boiled for ten minutes, then sliced into sections. These sections are fried in olive oil, seasoned with tamari or oil and apple cider vinegar or whatever you like, and bon appétit.

Carolyn also tells us of a method the Apaches liked especially well. After liberating the starch core and washing it, they would cook it in meat stews until it was tender.

The Yumans ate the starchy core raw or would break it up and dry it in the sun for short-term storage. When they were ready to use it, they would employ a mortar and pestle to crush it, and then it was boiled with fish.

Other tribes often added them to various kinds of soups. The Iroquois boiled the cores of the rhizomes down, down, down, until a sugary syrup resulted.

Eaten cold the flavor is significantly less appealing, so heat those leftovers up before serving.

Medicine:
Herbalists today are not likely to reach for cattails to remedy their clients' ills. However, First Nation peoples found them to be medicinally efficacious in a variety of ways.

Cattail Down

Cattail down was used by the Arikara, Blackfeet, and various California tribes as a soft, soothing dressing for burns, sores, and diaper rash. The Lakota mixed cattail down with coyote fat and used it as a salve to treat smallpox sores to prevent scarring.

Tribal women used the down as a menstrual pad that was particularly valued following childbirth. They also burned the down and used the ash to stop a newborn's navel from bleeding. A fluffy bed of down was made for mothers during childbirth, as well as for sick children. In cases of diarrhea, the mildly astringent down was consumed.

Pollen

In Chinese Medicine, the cattail pollen is used medicinally as a hemostatic herb that also promotes good circulation. The Chinese prescribe it for traumatic internal injuries that result in hemorrhages. It is also used to treat blood in the urine or stools, to stop coughing up or vomiting blood, to stop nose bleeds, and to dissolve blood clots.

The astringent pollen is toasted and taken as a tea to normalize excessive menstrual bleeding (menorrhagia) or for dysmenorrhea which makes menstruation difficult and painful due to severe congestion of the organs in the pelvic cavity.

Other conditions treated with the dried pollen include pain in the abdomen following childbirth, pain and pressure in the chest, chronic liver inflammation, and spermatorrhea, an abnormally frequent loss of semen without orgasm. It also serves as a diuretic.

Both the rhizomes and dried pollen are used to treat cystitis, a bladder inflammation caused by urinary tract infections. They are also used to treat hemorrhoids, drain pus from infected wounds, and treat urethritis, an inflammation of the urethra that is the result of a bacterial or viral infection or blockage (more common in males).

The recommended dosage for internal use is 4 to 9 grams of pollen twice daily on an empty stomach. Discontinue once symptoms are no longer present.

Rhizomes
California tribes beat the rhizomes into a jelly-like form and used it as a poultice for boils, burns, carbuncles, eye conditions, infections, inflammation, open wounds, scalds, sores, tumors, and ulcers.

In his excellent profile on cattails *Edible and Medicinal Plants of the Rocky Mountains and Neighboring Territories,* Terry Willard tells us that the rhizomes, when infused in milk, will alleviate dysentery and diarrhea. For treating kidney stones, a tea from the rhizomes of Typha angustifolia has sometimes been employed.

In his book *Tom Brown's Field Guide to Wilderness Survival,* tracker Tom Brown recommends that rhizome juice can serve as a satisfactory substitute for Novocain in dental extractions by rubbing it on the gums.

Stems
A concoction made from the stems was taken orally to treat venereal disease, diarrhea, or to kill intestinal worms.

Warnings and Considerations:
Cattails growing in polluted water or along highways absorb toxins as readily as useful nutrients and may be poisonous.

Wild Iris, which can be mistaken for cattails, sometimes grows alongside cattails. Iris leaves, stems, and rhizomes are poisonous. Be sure you are collecting only cattail rhizomes or shoots. Rhizomes harvested during or just after flowering should be dried or cooked before use to avoid the burning acridness of the raw rhizomes.

Agricultural Uses:
Insect Habitat
Cattails are important habitat for predatory insects that feast on Willamette mites, a serious pest of grapes. The moral is, if you are looking for a place to put a vineyard, cattails are the ideal neighbor.

Compost
The cattail's abundant production of bio-mass can be an excellent source of organic matter for compost. One can also chop the above ground parts and use them for a moisture retentive mulch to suppress weeds.

Other Uses:
The evolution of the human race undoubtedly owes a great deal to the remarkable cattail. Indeed, cattails once played a central role in many traditional economies worldwide. Although much overlooked in modern times, a cattail utilization revival may not be far off.

Archaic uses of cattails, in addition to food and medicine, are numerous. Even today, in far flung places on Earth, many of these prehistoric arts are still in practice, not to mention the new uses that have been discovered in the twentieth century. Furthermore, some cattail products are nothing more than refinements evolved from ancient crafts, including the making of small boats to ford storm-swelled rivers.

Waste Water and Sewage Treatment
In Colorado in the early 1990s several small municipalities developed wetland waste treatment systems using tens of

thousands of cattails as well as other wetland plants to process their waste streams. These wetland processing systems can save millions of dollars compared to a conventional system, with additional savings annually on chemicals. Colorado has also installed smaller wetland waste water treatment systems in some of the state's parks and rest stops, even at Shambhala Mountain Center, a Buddhist retreat.

New Mexico has also developed these wetland systems in small communities, subdivisions, schools, and some single family homes. They are attractive. They don't smell, and they are inviting to waterfowl. Best of all, the effluent they discharge can be well within water quality guidelines when properly sized and implemented.

The city of Arcata, California saved eight million dollars when they installed their ecological wetland sewage treatment plant in the 1980s. The cattail rich system doubles as a beautiful wildlife sanctuary. A by-product of this system is fed to the city's fish hatchery, saving money there. Sewage requires far more wetland acreage to process than waste water.

Basket Weaving

Archeological evidence shows that from about 100 BCE through 600 CE, now called the "Basket Maker Period," Puebloan basketry evolved into a very sophisticated art. Baskets were made in many sizes as well as shapes, and they combined skillful techniques with beautiful adornment. During this era that preceded pottery, many different plants were used for basket making. Combinations were common, and cattail fiber was often used. Another Pre-Columbian southwestern culture, the O'Odham people, were still producing baskets of a high level of artistic quality in historic times after the Pueblos had switched to pottery.

The cattail leaves were braided or the flower stalks were split and dried to make the fibers. Fibers of other plant species might be used for a twill weave. Unsealed baskets could be used to store miscellaneous items, or they could be used for straining liquids or sifting dry products. For watertight baskets, the heated pitch of conifers like Piñon pine was applied as a sealing compound. Other tribes used cattails for baskets as well. A popular application was to form the basket's foundation with the stems.

Woven Products

Many tribes used the leaves to make sleeping mats or in some cases floor mats for traditional shelters like the tipi, sweat lodge, and sundance lodge. Screens of plaited cattail leaves were used like room dividers. Water repellent capes were also woven from the leaves. As in old world cultures, the leaves were used for roof thatch, while the stems served as lath over the roof beams to help support the thatch. The long mature leaves and stalks were employed to weave bags and hats. Strong cords could also be braided from the leaf fiber.

Caulk

Various tribes utilized the leaves with their sheaths attached as a caulking for their dwellings or to seal canoes. The water resistant leaves have also been used to seal wooden barrels.

Caning

An age-old product still fabricated today from cattail leaves is rush or caning material for the seats of furniture, particularly chairs. Tightly braided strands of the leaves create a webbing that provides a comfortable seat. Although artificial rush from synthetics has been mass produced, they do not compare to the superior cattail rush in beauty, durability, or longevity.

Toward the end of the growing season, the craftsperson collects mature but still fresh leaves. Once gathered, they remove the mid-ribs, and the long leaves are bundled, tied, and hung on racks in the shade to dry. Once they have dried, they are soaked in water to make them soft and pliable again. The leaves are then separated into strands and woven tightly together to produce a strong cord ready for caning. (Be aware: paint or varnish can damage natural rush.)

Bedding
Cattail down has long been a craft material for numerous cultures around the world. Some tribes used it to make bedding, pillows, mattresses, or to line cradleboards for baby's comfort. The down has also been used for diapers and poultices.

Insulation
Cattail down possesses excellent heat and cold insulating properties and has been compressed into insulation boards. These boards also provide soundproofing and have been used in recording studios for that purpose. Native tribes often used the down to insulate their shelters.

In pioneer days, mountain men would stuff the down into their boots to keep their feet warm and to protect against frostbite.

The down makes a good, but lumpy stuffing material for jackets, comforters, and furniture. It is also very buoyant and has served as in-fill for life jackets, life preservers, and life rafts. Euell Gibbons stuffed the down into plastic bags and used it to insulate his freezer.

Fiber, Adhesive, and Drying Oil
In 1947 the Syracuse University Department of Plant Sciences

began researching cattail's economic potential. Leland Marsh headed the project, and the results of his efforts are quite remarkable. Oddly his work has not, for the most part, been practically applied, and despite this plant's multiple purposes, high productivity, and potential in the marketplace, its value remains virtually ignored (save for a handful of traditional people and wild food enthusiasts). This is doubly surprising when we consider that cattails have been central economically to so many traditional cultures worldwide. Indeed, whole civilizations arose around this plant.

Marsh recognized the tremendous bio-mass produced by cattail colonies and speculated that a useful fiber could be fabricated from this bio-mass. He experimented with various chemicals until he found a method to convert the flowers, stems, and leaves into soft fibered rayon and paper.

From the standpoint of reducing deforestation for paper and textile products, his research makes tremendous sense today. Marsh recognized that cattails grew on marginal agricultural lands that farmers often considered wasted space. In addition, cattails are highly productive and resist most pests and diseases. Their regenerative powers indicate perpetual harvests that might be guided sustainably with ecological wetland management. However, efficient methods of harvesting and processing were never developed and Marsh's inquiry was dropped. This may be just as well since mowing cattail wetlands with heavy equipment would probably degrade the habitat and undoubtedly impact waterfowl, other birds, fish, and so on through the food chain. From an ecological standpoint, commercial cattail harvests should rely primarily on low tech solutions and labor intensive methods.

In other experiments, Marsh fermented flour made from the rhizomes and produced ethyl alcohol which he envisioned as an industrial solvent and anti-freeze. Marsh also pressed the seeds to create a good quality drying oil. He fabricated a nutritious chicken feed from the by-product. From cattail stalks Marsh discovered he could also make an adhesive.

Candle Molds
Pioneers used cattail's strong hollow stalk as a candle mold. They simply threaded a wick through the hollow and dipped them in hot wax.

Miscellaneous Uses
The seed down was a popular tinder used for starting fires. In the nineteenth century, the stalk, topped with a mature flower cluster, was dipped in coal oil and used as a torch to light outdoor events.

Euell Gibbons once informed us how the stems can serve children's interests. The flower cluster is cut off, and the stems are sectioned into various uniform sizes of mini-logs for model log cabins and the like. Fabric pins are used to lock the logs in place.

Less functional than the applications discussed so far, but possibly of psychological benefit, is the popular use of the stems with mature clusters as a long lasting floral display. Indeed, since prehistoric times the decorative cattail has furnished many homes with an aesthetic counterpoint.

Ecological Functions:
Cattails are an inexpensive tool for restoring polluted riparian and wetland ecosystems. In today's toxified world, this attribute should not be understated. For example, cattails planted on the edge of creeks contaminated with nitrogen and phosphorous effluents can purify the water by 95% in a matter of weeks. One municipality purified their polluted streams for one million dollars with cattails rather than the

seven million dollars a high tech solution would have cost. This remarkable plant has demonstrated an ability to convert raw sewage to clean water and even lift heavy metals from mine tailings. From graywater to municipal waste treatment, cattails offer a natural solution.

Cattails also have value for flood control and stream bank stabilization.

Over time cattails are adept at transforming habitat from marsh into fertile meadow land, a significant ecological function in itself.

Cattails are of tremendous value as shelter and nesting for water fowl. Numerous bird species benefit and are often dependent on cattail colonies. For example, pheasants, various geese, ducks, blackbirds, marsh wrens, green-winged teal, and sora rails inhabit them. Geese are adroit at harvesting the rhizomes for food.

Rhizomes are also the preferred food of muskrats, who also use the stems and leaves to build their shelters. Muskrats have a habit of creating open water Cattail lagoons for attracting fish.

The young shoots are a gourmet dinner for elk, who feed on them avidly when they are available.

Native Range and Habitat:

Cattails require wet soil. The edge of wetlands where the soil becomes dry is the limit of their expansion. Ponds, lake shores, marshes, swamps, and along slow moving streams, wherever the soil is muddy or wet, they may call home.

Typha latifolia can be found outside Arctic climes from Alaska to Newfoundland and south throughout most of the continental United States from California to the east coast and into Mexico. The species is also indigenous to all of Europe (except the islands), Russia, and Asia. It is found from sea level to around 2,500 feet in Washington, below 6,000 feet in the Sierra-Nevada range, between 3,500 and 7,500 feet in Arizona's uplands, and from 4,000 to near 8,000 feet in Colorado.

Typha angustifolia is most common in coastal regions and is found in British Columbia, southeast Canada, and throughout much of the western United States. It also exists in the east from southern Maine to North Carolina and is a native of all Europe (except the islands).

Typha domingensis is native from Oregon to California and is found across the southern half of the United States down to South America's tropics, as well as in Europe and Asia. It is seldom found above 5,000 feet in the United States.

These three species readily hybridize and trihybridize where their range overlaps.

YUCCA SPECIES
[Agavaceae Family]

The uniquely formed yucca is often disdained by some because of their sharp needle-pointed leaves. In their natural range, they are common enough to be thought of as a weed, but with yucca one should not rush too quickly to judgment. Some Yucca species were a significant part of the Pueblo civilization's economy for 2,000 years and is still of religious significance in the Southwest today.

Yucca's common names abound. Often they are used generically for all yuccas. Some of these names are Our Lord's candle, Adam's needle, Spanish bayonet, soaproot, amole, and palmilla.

Yuccas are evergreen perennials that grow from a large, somewhat woody root. Many narrow, stiff, spine-tipped leaves radiate more or less symmetrically. The linear leaves are sometimes pale green or commonly bluish green, covered with a smooth wax. /In some instances, the leaf margin will have a thin whitish line down its entire length.

The common name, "Our Lord's candle," is a reference to yucca's beautiful floral display. Attractive bell flowers are borne copiously along tall flower stalks. The flowers are creamy white to greenish white, and some are fragrant, particularly in the evenings. The outer petals are often tinted a rosy color, while the inner petals sometimes display greenish highlights.

By day, the flowers of Yucca glauca and others are bell-

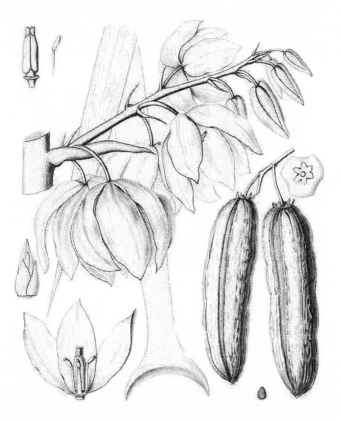

Yucca Aloifolia

shaped and droopy, but after nightfall, when the sky has plunged into darkness, they open wide. This is timed to summon yucca's small night flying pollinator, the yucca or pronumba moth (Pronumba yuccasella). This moth and the yucca's life are intimately entwined. They are dependent on each other to successfully complete their life cycles. Although the yucca can produce offspring by root offsets, it cannot develop viable seed without the aid of the yucca moth. In the course of pollinating the flowers, the moth lays her egg by boring a hole in the flower's ovary. But for the ovules to develop and supply the moth's larvae with food, the flower must be fertilized. The yucca moth appears well aware of this since she purposefully collects the flower's pollen and transfers it to the next flower, leaving enough ovules without eggs for an abundance of viable seed for her larvae.

Trunkless yucca's large thick roots are rhizomatous and form new crowns by offsets from the parent rhizome. These then form tightly clustered colonies of new yuccas.

Yucca in General
Culture:

Yuccas are sun loving plants, but a few species also do quite well in light or even partial shade. Too much shade, however, will inhibit flowering. Any well-drained soil should suffice. Sandy, coarse loams, and alluvial soils are ideal. Yuccas are more likely to be found in soils that are alkaline to very alkaline, but they tolerate slightly acidic soil. Yuccas are very tolerant of drought, but fail on ground that is too wet.

The seeds are quite viable if pollinated, and they will germinate easily. Germination in warm soils is necessary and typically occurs in seven to ten days. Sow seeds in the spring. When seedlings reach 6 to 8 inches, they can be readily transplanted. Offsets that are severed from the rhizomes with the leaves intact can be potted in soil. Offsets should be kept moist while the young root develops and becomes established. About a year after potting, the plant should be ready to set out in a permanent location. (See the profile of each species for cultural exceptions.)

Food:
Many western tribes were known to consume the young flower stalks, buds, flower petals, and immature fruit of the small, dry fruited species and the ripe fleshy, fruit of the broadleaf yuccas.

Flower Stalks
Before the flowers begin to bud, the freshly emerged flower stalks are rich in sugars and are delicious boiled or steamed until tender and are sometimes peeled. The stalks resemble giant asparagus spears, and in some Yucca species, taste vaguely similar, but much better.

Flower Buds
Another taste treat comes from the unopened flower buds. Once they swell out, they are eaten cooked. Due to their high sugar content, they can be delightfully sweet. Steam or boil fifteen to twenty minutes. They taste similar to green beans.

Flowers
The large flower petals also possess a nice flavor and texture, provided they are liberated from the bitter and somewhat tough flower center. Fresh petals can be added to salads or they may be steamed, boiled, roasted, or sautéed and served as a side vegetable or as a relish for other dishes. Flowers are steamed or boiled for fifteen to twenty minutes.

Tribes would often dry the petals for storage, then later

grind them with mortar and pestle so they could be used to season winter soups and stews. Early settlers pickled them.

Fruits

There are two types of edible yucca fruits. One kind is large and pulpy (like those of banana yucca). The other matures to a smaller, drier capsule (like those of soapweed yucca). Both are tasty once they have been properly prepared.

Carol Niethammer's approach to preparing banana yuccas goes something like this. The large, fleshy yucca fruits—found only on broadleaf yuccas—are typically boiled for up to half an hour. Then they are peeled, and the seeds are removed and set aside for other purposes. The pulp is mashed and pan-fried until it resembles jam, then a bit of sweetener such as coconut sugar, brown rice syrup, or honey can be added. It can be eaten as is, used as pie a filler, or mixed with a bit of whole flour and used as a filler in other baked goods.

To dry the pulp for future use, boil it down to a paste, roll it into sheets about an inch thick, and sun-dry it or slowly dry it in the oven at about 200° F. Once dried, it can be eaten as is or dissolved in water for a refreshing and nourishing drink.

The small, dry yucca fruits found on narrowleaf yuccas are harvested for food while still immature. They should be peeled and then steamed or boiled for fifteen to twenty-five minutes. Changing the water half way through the cooking process can help remove whatever bitterness still remains after the skin has been peeled. Season to taste, and they are similar to a savory, slightly sweet squash.

Seeds

The seeds of banana yucca are edible raw. They have a distinct licorice flavor, are chewy, and make a wholesome snack.

Roots
The roots are used as a foaming agent in root beer.

Medicine:
Functions
Yucca species have been used medicinally by numerous tribal groups in North America and have also been used in the folk medicine of European settlers. Most of the scientific research has come from three species—Yucca baccata, Yucca glauca, and Yucca schidigera.

Yucca is considered antiarthritic, antibacterial, anti-inflammatory, antimutagenic, antitumor, and a blood purifier. In addition, the saponins have antiviral and antifungal actions.

While yucca saponins inhibit some of the most harmful microbes, they simultaneously promote many that are highly beneficial, meaning the body's elimination systems do not have to work as hard to remove toxins. Yucca saponins may indirectly allow increased absorption of nutrients while obstructing the absorption of toxins. This may account for yucca's anti-inflammatory actions and why it offers arthritic relief.

There has long been a suspected link between some forms of arthritis and certain gastrointestinal tract difficulties. One school of thought suggests these stomach problems, and perhaps the arthritis itself, may be the result of stress on the beneficial intestinal flora, inhibiting their functions in the body. It is thought that yucca's steroid saponins have a positive rejuvenating and stimulating effect on these microbes, and act as a catalyst to relieve intestinal disorders and some cases of arthritis.

Roots
Numerous First Nation groups used yucca root medicinally.

The juices or suds of yucca roots were used to treat blindness and cataracts. A poultice of the crushed root was applied to the chest of those suffering from heatstroke.

Yucca roots' surfactant properties break down the fatty (lipid) envelope coating around certain viruses and microbial fungi, including the *candida* and *herpes* viruses. A tea made from yucca's inner root may also offer relief for inflamed colons, prostates, urethras, and joints.

Yucca roots are most often peeled and dried and may also be powdered. In his book *Medicinal Plants of the Mountain West*, Michael Moore says that for arthritics, a quarter ounce of the inner root should be boiled in a pint of water for fifteen minutes and taken three to four times daily. For more general purposes, research chemist Mark Pederson, author of *Nutritional Herbology,* recommends a 9 gram extract of the dried root with 45 milliliters of alcohol and 45 milliliters of water. In the case of capsules, capsules containing 490 milligrams are taken up to four times daily.

Fruit
The fruit was eaten by some southwestern tribes to ease childbirth.

Flowers:
Research with mice isolated an antitumor polysaccharide in a water extract of the fresh flowers of Yucca glauca that is active against B16 melanoma. Today, yucca supplies the pharmaceutical industry with many of its steroidal saponins.

Leaves
Some First Nation peoples made tea from the leaves to treat heartburn and vomiting. In the folklore of the early settlers of the Southwest, yucca was considered an anti-inflammatory for

the treatment of arthritis and rheumatism, as well as for diabetes and digestive problems.

Warnings and Considerations:
Yucca saponins have some toxicity. Very little research on its safe use is available. Michael Moore has warned that taking roots internally (even in the correct doses) can inhibit the small intestine's ability to absorb fat-soluble vitamins if used daily for long periods.

Agricultural Uses:
Wetting Agent
In the arid and semi-arid west, yucca extracts have proven invaluable agriculturally. As wetting agents, quick-acting surfactants in yucca extracts outperform comparable petroleum and alcohol based products. Many organic farmers can maintain an edge with this natural aid that is organically certifiable. When yucca extracts are added to irrigation water, the water's surface tension is reduced and water can penetrate deeper and more quickly into the soil. Yucca extracts will floculate heavy or compacted hydrophobic soils. In addition, yucca extract will also cleanse low-pressure irrigation systems, particularly the emitters. Yucca schidigera is most often used for commercial wetting agents.

Spreader/Sticker
When yucca extract is added to foliar sprays, the area the compost tea or fertilizer covers is increased and it sticks to the leaves longer, allowing more nutrients to be absorbed by the plant. This could also make dormant oil sprays and other pest control sprays more effective.

Anti-stress Factor

Yucca contains an anti-stress factor. Dry climate studies conducted on commercially grown alfalfa, broccoli, citrus and other orchard crops, cotton, onions, potatoes, strawberries, and tomatoes demonstrated 15 to 25% higher yields during periods of drought in fields inoculated with yucca extract over fields that were not treated with the extract. When the climate behaves ideally rather than harshly, little stress is generated and yucca provides no noticeable influence. But if the situation turns mean, yucca compounds are ready to act quickly to protect plants against adversity, particularly drought. They can also provide quick relief from winter freeze damage or from weltering heat, and is even effective on alkaline soils.

Living Barrier

Yucca can be a useful barrier for protecting gardens from some small predators.

Composting Agent

Yucca's remarkable recycling abilities can be a boon to composting operations. Yucca extracts provide a rich banquet of nutrients that yields an explosion of beneficial bacteria populations that can speed up the composting process considerably.

Pest Control

Yucca saponins have molluscidal properties. Agricultural pests like snails and slugs find these saponins to be highly toxic. For humans and other warm-blooded creatures, they are relatively innocuous.

Other Uses:

Soap

For most southwestern tribes, yucca was integral to their culture. Not only was yucca central to tribal economics, it was also of ritual significance in rites of purification. Once the Spanish invaders took up residence in tribal territories, they quickly learned the virtues of this plant. Yucca became popular with other early Euro-Americans as they too swarmed into the area. It was especially valued by the newcomers for soap.

When the Pueblo people temporarily expelled the Spanish conquistadors in the 17ᵗʰ century, they gathered together at rivers and streams to cleanse away the toxic effects of colonization and engaged in pagan baptisms with yucca soap.

Pueblo peoples typically washed and dried the whole root for future use. However, it dries more quickly if split lengthwise first. The root might also be used fresh. Before using, the roots themselves are washed in plain water to remove dirt and grit, then they are cut up into small pieces, retaining the bark. The pieces of root are placed on a stone and pounded with a wooden mallet to make them soft. The pieces are added to cold water where the suds are squeezed from them manually. Yucca lather quickly forms and the pulp residue is then skimmed off. Hot water can be added for washing fabrics or shampooing the hair.

Another method for producing a liquid soap involved slowly boiling the roots down until a storable soap concentrate resulted. The southwestern Hispanics much preferred yucca over other soap for cleaning wool since it left wool more soft and downy than even your local dry cleaners are capable today—and without manufacturing greenhouse gases and other toxins.

Today yucca is very popular with the natural product

industry. Its use in soaps, shampoos, perfumes, and cosmetics has inspired commercial cultivation on farms to meet the demand. Used primarily as a sudsing agent, it is also popular for bestowing body to hair and luster to dark hair. Yucca shampoo is reputed to fight dandruff and halt balding and even makes hair grow faster. In addition, it is a folk remedy for eradicating head lice.

The leaves of yucca also possess saponins, though not as richly as the roots. When native people would liberate the valuable fibers from the leaves they might use the leftover pulp for soap. Euro-Americans, on the other hand, found the leaf pulp nutritious as livestock feed.

Fiber

The fibers found in the yucca leaves have been used by First Nation peoples since prehistoric times for an incredible number of products. Once they had separated the fibers from the leaf and converted it to thread, twine, or rope, they would make clothing, blankets, sandals, hats, bags, paint brushes, baskets, brooms, belts, mats, caning, structural support for chairs and beds, fish nets, bow strings, tapestry, traps, scrubbers, and miscellaneous crafts and toys. Even their dwellings and ladders might be lashed together with yucca cords. Often the sharp needle on the tip of the leaf was retained with the fiber for sewing.

For two thousand years or longer, Pueblo peoples were making exquisitely crafted sandals and other items from these fibers. Sometimes they would weave wool together with yucca fiber when crafting various articles. Yucca paint brushes are still used today to paint Puebloan pottery.

In the 20th century the United States' government had cause to rely on yucca's remarkable fibers. During fiber shortages caused by both World Wars, our government utilized yucca fiber extensively for heavy paper, burlap, rope, and twine. In World War I alone, 80 million pounds of yucca fiber were used. In more recent times, a heavy craft paper for weather-stripping and flashing is made from yucca fiber.

To extract the fiber in the pre-Columbian manner, the leaves are first given a good soaking in water. Then the pulp is separated from the saturated leaves by beating them on a flat rock with a wood mallet, stone, or club. Intermittently they are dunked in water to help set the soft tissues free from the fiber. Lastly, they are cleaned with a scraper.

When collecting the yucca leaves for fiber, the youngest leaves are preferred, and summer is the desired harvest season. Once the leaves are cut away from the crown and the fiber is liberated, the fibers are separated by thickness and length in accordance with their intended use. Then they are dried in the sun until they fade to a soft off-white. This thread can be used as is for sewing or it can be woven into string, twine, cords, or rope. When making sandals, some tribes used the whole leaf or leaves that were split in half.

The O'odham peoples (erroneously called Papago and Pima) used the yucca to produce their flawless basketry. They also made a divining plaque from the fiber for resolving questions or predicting the future. These plaques were round in shape with the fibers spiraling out from the center. Occasionally new fibers were added on the journey to the outside edge. A loop was tied to the outside border so the plaque could be hung on the wall. To receive an answer to a question, one selected a thread from the outer edge and traced its course toward the center. If the thread reached all the way to the center the outcome looked good. The shorter the chosen thread, the less favorable would be the result.

Sewage Treatment

Yucca extract has also found a niche in municipal sewage and waste treatment facilities. Research shows that the extract dramatically stimulates beneficial microbial activity and quickly leads to striking microbe population increases. These microscopic aerobic communities devour organic waste and greatly accelerate its decomposition. Obviously this effect serves the goals of reducing sewage build up and diminishing the concentration of toxins in the water.

Firesafe

Yuccas are very fire resistant, particularly those without trunks. In wildfire prone areas, they are recommended as ornamentals for firesafe landscapes.

Odor Control

Yucca schidigera is used as an additive in dog and cat food to effectively control the unpleasant stench of the animals' feces.

Jewelry

The seeds of some Yucca species are attractive enough to use in jewelry making. For example, the seeds of Yucca baccata are triangular and flat with slightly rounded corners. They are a dull black color and all the edges are embossed in a repetitive pattern. They are excellent for necklaces, wristbands, earrings, etc.

Religious Ceremonies

While the first Hispanics to arrive in the Americas learned to use the soapweed yucca for medicinal uses, they put it to other uses as well. Perhaps the most unique way it was employed was by the secret Hermanos Penitentes religion which is still practiced in northern New Mexico and southern

Colorado. The Penitentes purposefully take on the burden of Christ and voluntarily ritualize his suffering. They use yucca to make a flail so they might experience Christ's pain under the lash. First they drink a beverage brewed from the young shoots of yucca in order to gain the strength and courage to face the coming pain of the whip and the agony of crucifixion. Though obviously a severe movement, the Hermanos Penitentes' rituals do seem to generate humility and compassion in some of its practitioners.

Ecological Functions:
Keystone Species
In its native habitat, yucca may be a keystone species, meaning they may play a central role in the healthful dynamics of their eco-system and its sustainability. As their leaves and other parts age and die, they begin to break down and their organic debris is blown about. Once this organic debris has settled, it begins to decompose into the soil and release steroidal saponins. These are complex sugars that help plants assimilate nutrients and improve their resistance to stress. The roots of other native plants will then take these steroidal saponins into their immune systems. In addition to increased nutrients and stress resistance, plants also become endowed with an increased ability to utilize available moisture and thus have greater drought tolerance.

Yucca steroidal saponins also help plants because their decomposing organic matter stimulates vital microbial communities.

Water and Soil Rehabilitation
Yucca extract can cleanse and clarify water polluted by mineral salts, as well as rehabilitate (flocculate) fragile soils that have biologically and structurally collapsed and no longer

support life. These soils are typically heavy clays that were subjected to over-irrigation, resulting in heavy salt deposition in arid and semi-arid landscapes.

Wildlife Habitat and Livestock Fodder
The tender young leaves of immature plants, the young flower stalks, and the flowers and fruit are consumed by deer. Birds and insects dine on the fruit. Cattle and sheep also feed on yucca. A few bird species, like orioles, that may have difficulty finding a proper nesting site in the more arid regions, will often nest in the protection of the yucca's spiny armament.

Native Range and Habitat:
The individual profiles of each species contain specific ranges and habitats.

Species:
Of the forty or so Yuccas native to North America, the following Yucca species, all natives, have the most documentation on their traditional and contemporary uses in the continental United States.

Y. Aloifolia
(Spanish Bayonet)

Slow growing to about 10 feet high and 5 feet wide, this treelike yucca may reach 25 feet in great old age. Its leaves are about 2½ feet long and 2½ inches wide with flower stems up to 2 feet tall. This yucca commonly grows on sand dunes. Its fruit pulp is used as food, and its leaves provide fiber. It is native mostly in coastal areas from Virginia south to Florida and west to Louisiana, and has naturalized along the Texas coast and parts of Florida. It is also a native of Mexico and the West Indies. Spanish bayonet is hardy in zone 8 to the tropics.

Y. Angustissima
(Fine Leaf Yucca)

Fine leaf yucca is typically a trunkless rosette with leaves 1½ feet long and ¼ inch wide. It has white flowers with greenish or purplish tints. Its flower stems are 2½ feet to nearly 7 feet tall. The dry pods are eaten after being thoroughly dried or roasted. They are also used for soap and fiber. It is a desert plant also found on dry mesas and the slopes of hills. They are native from southern Nevada east to Colorado and south into central Arizona and northwest New Mexico.

Y. Baccata
(Banana Yucca, Datil)

This yucca forms a trunkless rosette with leaves that are 28 inches long and 2½ inches wide, with flower stems shorter than most other yuccas. It produces a fleshy fruit up to 9 inches long, making it the second largest edible fruit native to the United States after pawpaws (Asimina triloba). It is also an important soap and fiber plant with dye and medicinal uses. It is commonly found with piñon/juniper. Elk are particularly fond of consuming the heart of this plant. It is native from southeast California to southwest Colorado, in southern Nevada and Utah, and south into north and central Arizona, and east to southwest Texas, commonly at altitudes from 4,000 and 7,000 feet. It is hardy to zone 5 and borderline in zone 4.

Y. Brevifolia
(Joshua Tree)

This is another treelike yucca that grows from 15 to 40 feet tall with a stout trunk of up to 3½ feet in diameter. Its leaves are about 14 inches long and ⅜ to ⅝ inches wide, and its flower stems are 8 to 20 inches tall. Its fleshy edible fruits grow up to 4 inches long and are quite plump. It is an important fiber plant. An upland desert plant, Joshua tree grows at altitudes of 2,000 feet to 6,500 feet and occasionally higher on arid, gravelly slopes. It is native from southeast California to southern Nevada, southwest Utah, and northwest Arizona. It is hardy to zone 8, but has been known to survive brief minus 20° F freezes.

Y. Elata
(Soaptree Yucca, Palmella)

This treelike yucca is commonly 6 to 15 feet tall, but in great old age may reach 30 feet in height. It can grow moderately or even fairly fast when young, but eventually becomes slow or even very slow growing. Yucca elata and possibly other yuccas with trunks will re-sprout (coppice) if cut down. Its leaves grow up to 38 inches long by 1 inch wide. Its flower stalks can be 3 feet to 10 feet tall. They are important fiber plants. In World War I they were utilized as a jute substitute for the manufacture of bagging. Its tough, but lightweight, flexible wood is used for medical splints. Although soaptree yucca prefers sandy or gravelly soil, it adapts to clay. It is an associate of mesquite. A plant of the desert and grasslands, soaptree yucca is native to Utah, south to southern Arizona, east to southwest Texas, and down into Mexico. Its range of altitude starts at 1,500 feet and climbs to 8,000 feet. It is hardy to zone 8 and borderline in zone 7.

Y. Filamentosa
(Adams Needle, Needle Palm)

Needle palm is a trunkless (or nearly so) rosette with 12 to 32-inch leaves that are 1 to 4 inches wide and 3 to 6 foot tall flower stalks. It is very soil adaptable as long as the soil is not too wet. It is a useful soap plant. Both the Cherokee and Catawba peoples used it for medicine. It is a plant of dry, open hardwood woodlands, and native from Maryland to Florida and west to Mississippi. It is hardy to zone 5 and borderline in zone 4 where it needs some protection from the cold.

Y. Glauca
[Syn. Y. Angustifolia]
(Soapweed Yucca, Narrowleaf Yucca, Great Plains Yucca)

Soapweed yucca is a trunkless rosette with leaves up to 28 inches long and ½ an inch wide. Its flower stalks are 2 to 6 feet tall. All yuccas tend to be uniquely attractive, but trunkless, narrow leaf species like soapweed are very beautiful in flower. It is one of the most heavily utilized yuccas for medicine and soap products. In World War I 80 million pounds of its multi-use fiber was harvested to produce burlap. In World War II its fiber was harvested to manufacture paper. It is native through the Great Basin from Montana to western North Dakota, south to Arizona and eastern New Mexico and northern Texas. It is hardy to zone 4.

Y. Shidigera
[Syn. Y. Mohavensis]
(Mojave Yucca, Spanish Dagger)

This shrubby yucca is 3 to 15 feet tall with leaves 1 to 4 feet long and 1 to 2 inches wide. Its flower stems are 6 feet long or longer. It is the most utilized yucca for commercial wetting agents and is used in today's soap and cosmetic industries as a mild detergent, a deodorant, a preservative, and a sudsing agent. It was also an important fiber plant during World War II. Its edible, fleshy fruits grow up to 4 inches long. Spanish Dagger has antimutagenic properties and is used as an herbal medicine. It is native from southern California to Nevada and east through Arizona and south into Mexico's Baja California. It prefers more moisture than most yucca plants, although it is a desert plant, often found on dry, rocky slopes at altitudes of less than 5,000 feet. It is hardy to zone 7.

Y. Whipplei
(Our Lord's Candle)

A trunkless rosette with leaves 1 to 2 feet long and ¾ inches wide. Its flower stems grow 6 to 14 feet tall. It is a strong fiber plant of industrial quality and is rich in cellulose. As an oilseed plant, it yields a protein rich oil. The dried flower stems have been used as surgical splints to hold dressings on fractures. It can be used for soaps as well. The plant dies after flowering, but tends to leave a cluster of progeny behind (offsets). It is native to the central California coast and the southern California Coast Range and its interior mountainous regions, and south into Mexico's Baja California. It is hardy to zone 8.

ZIZIPHUS JUJUBA
(Chinese Jujube, Jujube, Chinese Dates, Da-Zoa, Ta-Tsfootao)
[Rhamnaceae Family]

The jujube is a slow to moderate growing (about 1 foot per year), large deciduous shrub or small tree, usually 15 to 30 feet in height and 15 or more feet in width. Occasionally they can reach a height of 40 feet and, rarely, in great old age, 50 feet. The flowers are an inconspicuous whitish or yellowish green, appearing in clusters of six or more, with each flower about ¼ inch in diameter. The flowers may vary from very to mildly fragrant and are produced abundantly over a long period of time, often starting early in the summer and continuing well into late summer. The handsome, glossy dark green leaves are somewhat oval, but vary in shape. They are 1 to 3 inches long and turn a fine yellow in fall. With age, the trunks are short and gnarled, about 10 inches in diameter, with mottled gray to blackish bark that peels in large strips. The branches are slender, drooping, and develop in a unique zigzag pattern. The wild plants are quite thorny and tend to have a pair of slender 1 inch thorns at the base of each leaf. One is straight and one is curved. With age, the thorniness is much reduced, and often older trees are thorn free. Many of the cultivars are thornless, or nearly so. The egg- or pear-shaped fruits are a shiny mahogany red, reddish brown, or just plain brown. They are smooth at first but wrinkle when fully ripe, and

Ziziphus Jujuba

contain a single bony, two-seeded stone. The fruits ripen between September 1 and November 1; however, the fruits do not ripen all at once. In the wild, the fruits are typically ½ to 1 inch in diameter, but cultivars can produce fruit up to 2 inches in diameter.

The Chinese have cultivated jujubes for at least 4,000 years, and have developed as many as 400 cultivars. Jujube is probably grown more than any other fruit in China, and is cultivated throughout the country.

The ancient Greeks and Romans introduced the jujube to southern Europe, western Asia, and North Africa. It is cultivated today in Europe, South Asia, the Middle East, Africa, and Australia. It was introduced to the United States around the end of the 19th or beginning of the 20th century.

Culture:

Jujube is a very tough and adaptable plant, particularly in arid and semi-arid climates. It tolerates temperature extremes as severe as a blistering 120° F to winter temperatures down to minus 22° F. The plant survives in climates where annual rainfall reaches 80 inches or just a mere 5 inches. The jujube is reputed to perform well in western New York and southeast Pennsylvania. It has been known to thrive at 5,500 feet in southwest New Mexico and northeast to the outer foothills of the Ortiz Mountains between Santa Fe and Albuquerque. Jujubes have grown vigorously in western Colorado, from Montezuma in the south and Grand Junction in the north. It has been cultivated throughout much of California, Arizona, and in southern and eastern Oregon.

In the coldest part of its range, jujube performs best in hot sunny microclimates, and young plantings should be protected against the cold until they are 3 to 4 feet tall.

Jujubes are happy and productive where the growing season is long and hot, but do not grow as well in climatic situations where the seasons are short and cool. Fruit quality and yields will also suffer where protracted humidity occurs. Rain during the fruiting stage can split the fruit.

Jujube likes full sun, dislikes even modest shade. It is at its best in deep sandy loams and well-drained soil. It is very tolerant of strongly alkaline or highly saline soils. It doesn't tolerate much acidity. Jujube does well in nutrient deficient soil, but it is not particularly fond of clay. In general, however, it adapts to many soils.

Jujubes are quite drought-tolerant, thanks to their deep taproot which, unlike the above ground plant, grows fast. They are also reputed to withstand periodic flooding and poor drainage, but if the soil is soggy most of the time, it will suffer significantly. Periodic deep irrigation during fruiting season will create moister fruit and more of it.

Fertilization, in most cases, is not necessary or recommended.

Jujubes are nearly immune to pests and diseases. The only serious problem they might encounter is Texas root rot, a parasitic fungal disease common in the arid lowlands of the southwestern United States that is fatal to a large number of plant species.

Pruning is not strictly needed; however, young trees will benefit from training on a strong open framework. (They yield equally well whether pruned or not.) The species and cultivars will sucker freely when the soil is disturbed. In the case of grafted trees, thorny suckers are produced from the rootstock, sometimes well out from the trunk. These suckers should be removed regularly. Even cultivars on their own roots can be thorny and sucker; others, however, they may be thornless.

Propagation by seed usually produces thorny plants with poor fruit characteristics. Root cuttings or suckers are used to propagate cultivars that are not grafted. Clones taken from varieties growing on their own roots will not develop the characteristic tap root, and so will be more fussy about soil quality and moisture. Stem cuttings are difficult to root, but softwood cuttings will root easily in the proper setting. Grafting is the best way to clone cultivars that are as tough and adaptable as the species.

Pollination is by honey bees and other insects. Varieties require cross-pollination with other varieties for good fruit yields.

Varieties:

The Chinese have developed hundreds of varieties of jujube. Although only a handful can be found in the United States, more varieties are becoming available. Jujube seedling are inferior when compared to its fruiting cultivars. Some cultivars are self-fertile, however yields increase significantly if cross-pollinated by other varieties. Cultivars may produce some fruit their first year in the ground, but usually come into full bearing in a few years. They can require close attention and care to get established. Some varieties are thornless.

Ant Admire: This cultivar produces a long, narrow medium sized fruit with a very sweet flavor. It ripens in midseason.

Black Sea: Fruit has a very good flavor, and it is very productive. It was discovered in Yalta on the Crimean peninsula.

Chico: This cultivar's fruit is medium size. Many of its smaller fruits are seedless. The mildly acidic flavor is unique, and the flesh of the fresh fruit is quite crisp. Chico is very thorny. It has done very well in Southern California.

Coco: The coco's fruit flavor is unique, with a distinctive coconut flavor. It was also discovered in Yalta.

Contorted Jujube: Contorted jujube is small fruited, but that fruit is quite tasty. The trunk and branches are seriously gnarled, giving the tree an ancient look. It can be used as a pollinator or for its unique wood.

GA 866: This cultivar produces a 2-inch long fruit that is very sweet and reads 45% Brix soluble sugar.

Honey Jar: Honey jar produces fruit that is smaller (⅜ to 1 inch) than any of the other cultivars, but it is extremely sweet whether fresh or dried, and is exceptionally delicious.

Intermis: This is a thornless variety that might make good rootstock.

Jin: Jin produces fruit that is 2 inches or more in length. It is delightfully delicious fresh or dried. The dried fruit tastes more date-like than other cultivars.

Lang: This cultivar produces pear-shaped fruit, ranging from 1½ to 2 inches long, and ½ to ¾ inch wide, but the smaller size may be more common. It is sweetly flavorful with melting flesh, extremely productive, and very tasty fresh, dried, or candied. It is a large, thornless, upright attractive tree, most commonly grown in the western United States.

Leon Burk Bellflower: The fruit of this cultivar is possibly the juiciest of all, is highly productive, regularly bearing, and somewhat self-fertile. It is widely grown in southern Georgia, but has not been available in California.

Li: Li produces one of the largest fruits among the cultivars, commonly reaching 2 inches long. It contains a tiny seed, has excellent flavor, and is quite sweet with a crisp flesh. It is high-yielding, self-fertile, and ripens in late August. The tree is low growing, but wide spreading, thornless, and has glossy, particularly attractive foliage. It is commonly grown in the West.

Ming Tsao: Ming Tsao produces a sweet, 1½ inch long fruit. It is an upright tree that is small in stature and thornless or nearly so.

Mu Shing Hong Tsao: This cultivar produces a medium-sized fruit that is occasionally seedless. It is nearly thornless and produces few suckers.

Qiyue Xian: The Qiyue Xian has sweet, medium size fruit that give good yields. It is more cold tolerant than many other cultivars. It needs a pollinator.

Redlands 4: Redlands 4 produces large fruit, even a bit bigger than Li. It has a superb sweet taste and crisp texture. It ripens late.

Sherwood: This cultivar produces a fairly large fruit, 1½ to 2 inches long, of superior quality. It is one of the very best, is extremely productive, and has fairly small thorns, much fewer than most. It is commonly grown in the southeastern United States.

Shui Men: Shui Men produces medium-sized fruit that is delicious and of excellent quality fresh or dry. It is a productive variety.

Silverhill: Silverhill produces a plum-shaped fruit, 1½ to 2 inches long. It has a sweet, very tasty date-like flavor when dried, and is of excellent quality, both fresh and dried. The tree is large with few suckers or thorns.

Silverhill Round: This cultivar is very similar to Silverhill, but with smaller fruit 1 to 1½ inches long. When first ripe, but still hard, the fruit is unusually rich in sugars. Once it becomes soft, ripe, and wrinkled, the flavor resembles that of a prune. It ripens in September.

So: So produces a moderate-sized fruit of good quality with moderate yields. It ripens early and is a slow growing genetic dwarf. The trunk becomes very gnarled with age.

Sugar Cane: Sugar Cane produces a small fruit about 1 inch long. It is exceptionally sweet and crunchy when eaten fresh. It ripens mid-season and needs a pollinator. The tree is very thorny.

Twen Ku Lu Tsao: The fruit is exceptionally, sweet and is a good keeper, but the yields are low. It has few suckers.

Wuhu Tsao: Although not readily available, this variety is mentioned for its potential to contribute to the development of improved cultivars. The fruit is very sweet and the stone is soft and edible, so pitting is not required.

Food:
Fruit

Jujube fruits are very rich in vitamin C, a source of vitamin A, some B vitamins. They contain roughly about 22% sugar, plus various organic acids, pectin, and mucilage.

Between the time the green fruit begins to color, but just before it starts to wrinkle (a period of 3 to 5 days), the fruit is crisp and firm. It can be harvested during this period and eaten fresh. It tastes more or less like an apple. Left on the tree to fully color and wrinkle, it develops a gummy texture similar to a date, and it is often harvested at this stage.

Jujube fruits are eaten fresh or sun dried, and have been canned, stewed, baked, pickled, brandied, smoked, jellied, or used for beverages. They are cooked with grains, used for pudding, or added to breads, cakes, or soups. The fruit is popular for candies. They can be pricked all over, allowing the juices to ooze out and crystallize on the fruit's surface, making it a delightful sweet treat. In India, the seeds are pressed to yield an edible oil.

Jujubes are very fruitful. A single mature tree can yield 60 to 100 pounds of fruit. The fruit keeps well after a short period of drying in the sun. It stores well under refrigeration as well, but deteriorates quickly if left on the tree. Typically harvested in mid to late September, jujube will not ripen if picked green.

Medicine:

It has been a hundred years or more since jujubes were introduced in the United States, yet their tasty edible fruit and their medicinal virtues remain obscure in the states. In China we see the complete opposite. Jujube's fruit is very popular, and it has been an effective medicinal herb for millennia.

Most of what we know of jujube's medicinal applications comes from TCM. Jujube's fruit and seeds are both listed in the *Pharmacopoeia of the People's Republic of China* as official drugs with quite similar effects as those touted by TCM. It is highly regarded as a general tonic as well as a specific tonic for the brain, gallbladder, heart, liver, lungs, or stomach. It is also added to other potent tonics because it can increase the time they remain active.

Nervous Disorders

Jujube has a particularly beneficial effect on the nervous system as a mild but very effective sedative and nervine. It is an excellent treatment for neurasthenia (nervous exhaustion) and is a soothing relaxant for both body and mind. Animal studies have confirmed the fruit's effectiveness for the treatment of anxiety-induced insomnia.

As a hypnotic, jujube stimulates activity in the central nervous system that can result in a sleep-like state. It has been prescribed for hysteria, and the seeds are used as a mild liver sedative. It also provides relief for those suffering from fatigue.

Circulatory System

While jujube fruit is a circulatory system stimulant, that is not the whole story. The fruit also stimulates natural killer (NK) cells to roam the bloodstreams. These NK white blood cells are essential in fighting serious diseases such as cancer and AIDS, as they destroy harmful aberrant cells and prevent their reproduction.

Jujube is also combined with other herbs in TCM formulae for specific medical disorders. For example, jujube is a component in a formula designed to treat the potentially fatal condition idiopathic thrombocytopenic purpura (ITP), an autoimmune hemorrhagic condition where platelets are destroyed entering the spleen causing mucus membrane capillaries to produce small hemorrhages that leave purple rashes on the skin. In western medicine it is common to remove the spleen (the largest endocrine gland) which results in 50 to 60% remission.

At the Shanghai Medical University Children's Hospital a clinical study of ITP was conducted. Forty-one children suffering from ITP were selected for the study. An herbal formula developed to treat ITP that included jujube was used to treat these children for a period of approximately five months. All but one child improved, and there were twenty-four remissions—a rate of 97.6% effectiveness. They also got to keep that important lymphoid organ, the spleen (Allen K. Tillotson, *The One Earth Herbal Source Book*).

In another TCM herbal formula that includes jujube fruit, fourteen chronically ill children were treated for hepatitis. In each case the hepatitis B antigen was removed from their blood (Allen K. Tillotson, *The One Earth Herbal Source Book*).

A cardio-tonic made from the jujube root stimulates blood circulation and helps to normalize heart palpitations when treating strokes. Research confirms that the fruits and seeds of jujube can reduce high blood pressure by hypotensive action.

Jujube bark is an anti-hemorrhagic herb. It reduces or prevents hemorrhages and bleeding in general. It is used to treat patients who are vomiting blood and women suffering serious uterine bleeding. It is also a good regulator of menstruation. In these last treatments the bark is used. The bark is prepared by first reducing it to charcoal and then grinding it into a powder.

Cancer Treatment

Jujube contains the antibiotic ampicillin. For this reason, it was added to a TCM herbal formula to test its ability to fight lung cancer. Rats were given chemicals intended to result in lung cancer. Once the test was completed, it was found that the development of cancer cells was totally inhibited in the rats. More research is needed, and soon (Allen K. Tillotson, *The One Earth Herbal Source Book*).

Respiratory System

As a life-supporting lung tonic that promotes and nourishes

lung functions, jujube can play an important role in those suffering from asthma, chronic bronchitis, or pulmonary tuberculosis. Although the latter typically affects the lungs, it can spread to the gastrointestinal and genitourinary tracts and negatively affects the nervous system, lymph nodes, skin, bones, and joints.

Jujube, in addition to directly suppressing pulmonary tuberculosis, also helps to contain its spread. The protection is due to its effectiveness as a stomach tonic, its ability to nourish the abdominal organs, its ability to help the circulatory system to keep the joints in good condition, and its ability to strengthen the bones and sinew.

Miscellaneous

The fruit is used to reduce profuse sweating and night sweats, while also relieving chronic thirst. At the same time it acts as an antisclerotic, keeping the body's tissues appropriately hydrated. The seeds work as an antipyretic to reduce fever and are particularly popular for treating childhood fevers brought on by changing weather patterns. The root is used for urticaria (rashes or hives) and other inflammatory diseases of the skin.

Jujube is useful in nourishing the viscera (abdominal organs), as well as the lungs, and in treating malnutrition. It is a gentle but effective laxative, acts as a diuretic, and treats diarrhea in cases like dysentery. It serves the mucus membranes by acting as an expectorant, drying up or reducing excessive mucus.

A stomach tonic can be extracted from the fruit. Research with animals appears to validate this use by TCM doctors. It tones and relaxes the smooth muscle, and as an analgesic herb, it suppresses stomach aches and treats joint pain. An extract of jujube alcohol is reputed to have antiulcer properties.

The leaves are used as a vulnerary herb for healing wounds, sores, burns, boils, etc. It can be used to treat foul wounds that refuse to heal. A decoction can be used as a wash or a poultice of leaves can be applied.

In the East, jujube is considered a longevity herb that is also strengthening. Be aware that long-term consumption of jujube could make you younger and healthier.

Warnings and Considerations:

Jujube is non-toxic and the only contraindication to this is found in TCM. The Chinese recommend against jujube fruit for medicine when treating those afflicted with fullness of the mid-abdomen when copious phlegm is present.

Agricultural Uses:

Fodder

Jujube leaves serve as livestock fodder and are also used to feed the tasar silkworm, as well as a lac-producing insect. (Lac is a resinous secretion used to make shellac)

Living Fence

Jujube is a popular hedgerow plant and is planted as a shelterbelt component. The thorny cultivars are grown as livestock barriers to protect crops and can also provide forage and fruit. It is also recommended as a warm season windbreak component for further crop protection.

Bee Forage

Honey bees cannot resist the ambrosial fragrance of the flowers, which they pollinate and turn into an exquisite gourmet honey, providing there is a sufficient number of jujubes in the area.

Other Uses:
Wood Products
Jujube wood is tough, hard, strong, and durable. It takes a fine polish, and is also popular as a fuel wood. It is used in India for the manufacture of various articles, ranging from turnery to sandals, and tool handles to golf clubs.

Tanning and Dye
The fruit is employed in dying silk, and the bark, which is high in tannins (4 to 9%), is used for tanning leather.

Skin Care
Long term consumption of the fruits is believed to result in a beautiful, youthful complexion.

Native Range and Habitat:
The jujube is native to China, India, Pakistan, and southeast Europe. It has been cultivated for thousands of years and widely naturalized, making it unclear what its homeland parameters once were. However, jujube is thought by some to have originally migrated from Syria. Occasionally it naturalizes in parts of the United States.

In its native range, jujube is found in a wide range of environments. It may colonize riparian zones or be found on dry plains and foothills.

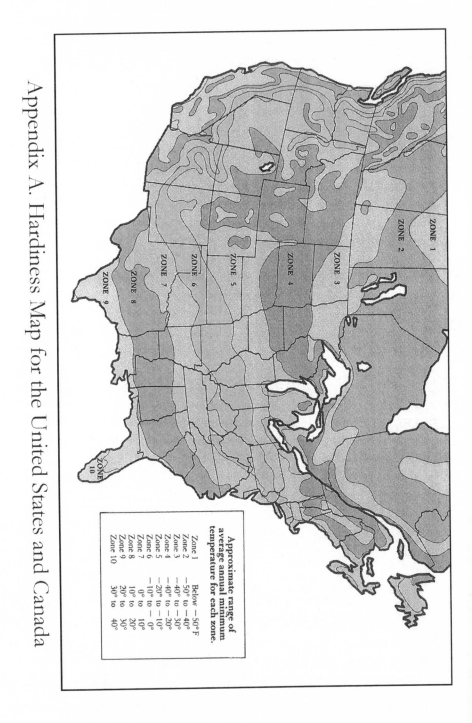

Appendix A. Hardiness Map for the United States and Canada

ZONE 1
ZONE 2
ZONE 3
ZONE 4
ZONE 5
ZONE 6
ZONE 7
ZONE 8
ZONE 9
ZONE 10

Approximate range of average annual minimum temperature for each zone.

Zone 1	Below −50° F
Zone 2	−50° to −40°
Zone 3	−40° to −30°
Zone 4	−30° to −20°
Zone 5	−20° to −10°
Zone 6	−10° to − 0°
Zone 7	0° to 10°
Zone 8	10° to 20°
Zone 9	20° to 30°
Zone 10	30° to 40°

Appendix B. Ecological Plant Culture: A Necessary Change

The design and culture of botanical treasure plantings is very important if they are going to help mitigate climate change and other environmental problems and reduce poverty. It is becoming increasingly important that permaculture design and agroecology quickly replace industrial agriculture, which has become a blight upon our planet.

Industrial agriculture is releasing about 750 million tons of CO_2 into the atmosphere annually while consuming about 40% of the United States energy stores. Organic agriculture, by contrast, can sequester about $3\frac{1}{2}$ tons of CO_2 per acre and is much more efficient in energy usage. Small organic farms can average as much as 200% more productivity than industrial agriculture, and when a small village collectively practices organic agriculture, they have been shown to be as much as 400% more productive per acre than industrial farming.

In 2006 an agricultural census conducted by the United States government reported that small farms produce more food per acre—whether measured in tons, calories, or dollars—than large industrial farms. In addition, small farms are more efficient in their use of land, water, and petroleum products.

Industrial farming is heavily invested in the use of toxic inputs that destroy living soils and pollute water, air, and the food itself. This is not the case with agroecology, permaculture, and conventional organic farming.

Because of this dependence on fossil fuels, the cost of food will continue to rise as crude oil becomes increasingly expensive to extract. This will predictably lead to more hunger, food deserts, and rising costs for most commodities.

Permaculture/Agroecology Methods in Brief

Site Investigation:

Know your land before you start. Knowledge of on site soils, water availability, annual precipitation, problem and prevailing winds, topography, orientation, thermal belts, draws, and existing flora is essential to the design of productive and sustainable agricultural systems. Learn to quiet your mind and listen to the land that speaks in impressions rather than words and it will help guide you.

Ecological Design:

Ecology is the foundation of appropriate design. Designing agricultural systems ecologically can only be hinted at here, but the knowledge gained from a site investigation is the beginning of intimacy with the land and is essential for good design. The goal of ecological designs is to create land use that is co-evolutionary with the natural system. Ecological design for agriculture emphasizes synergistic, biodiverse communities of flora and fauna that interact mutually. In essence, a renewable self-sustaining system is the ideal.

Bill Mollison, author of *Permaculture: A Designer's Manual* and the force behind the permaculture design movement, stresses the importance of values and ethics. A good permaculture designer has a deep commitment to care for our earth as well as the wellbeing of our fellow human beings. Good design should provide for the basic needs of the greatest number of people within the ecological capacity of the site.

Terraforming:

Terracing is one of many forms of terraforming. Terraforming is about reshaping the land to make the best use of precipitation. It has sometimes been used to make it easier to tend to crops or to control pests. Swales, pit catchments, and some kinds of china'mpas are other types of terraforming.

Soil:

The soil preference of botanical treasures can vary widely by species. Some species do well on marginal soils but may do better on more fertile soil. Others may require rich soil and still others require poor soils. All soils that support plants without artificial chemical fertilizers are, in general, biologically alive.

The richer the soil the greater the biological community it supports. A vibrant microbial community is essential, as are abundant organic matter and mycorrhizal fungi. Soil can be a very complex subject to go into, but increasing natural soil fertility is of critical importance.

Two ways that soils are enriched are through composting and cover-cropping (commonly using plants associated with nitrogen fixation), also known as green-manuring. Both add nutrients and organic matter (carbon) to the soil. In permanent ground covers, mineral hyperaccumulators growing with perennials (particularly nitrogen fixers) are best used in orchards and shelter belts.

Research has long shown that the excessive use of both organic and inorganic nitrogen fertilizers dramatically reduced the vitamin C content in food crops. It also impacted the ability of soil microbes to consume methane and can contaminate forest ecosystems and water courses by drift. This is unlikely to happen when using nitrogen fixing plants.

An ancient method of soil amendment is through *agrichar* or *biochar*, a product made by cooking wood down to charcoal by pyrolysis. This charcoal makes a powerful soil amendment for enriching poor soils.

Global warming is changing the chemistry of our soils through the increased buildup of carbon dioxide in the atmosphere. We will need to learn how to adapt our soil stewardship to address these changes as they become more and more evident.

Water:

Depending on local climate conditions, rain, snow, fog, or dew can be captured for multiple uses, including irrigation. As stated previously, terraforming can in a variety of situations dramatically increase the capture and use of precipitation.

In the draws of hill sides, that is the saddle between two hills where runoff accumulates due to gravity, a series of leaky check dams from top to bottom can capture and filter runoff. In arid and semi-arid lands, moisture lost to evaporation can be radically reduced in this way. In areas where precipitation is much higher, the check dams reduce flooding, and in all climates, can prompt the emergence of springs in the lower areas. The dams themselves are excellent planting sites because they capture eroding topsoil and concentrate moisture.

Precipitation falling on roofs can, if harvested, provide a great deal of water. In general, about 600 gallons of rainwater can be collected from 1,000 square feet of sealed roof for each inch of rainfall. The actual amount will very somewhat depending on the type of roofing (any sealed surface will work) or the efficiency of the rainwater harvesting system. Around 6,000 gallons can be harvested per 1,000 square feet of collection area in arid lands that receive 10 inches of annual rainfall. Where annual rainfall is more like 50 inches per year, about 30,000 gallons per 1,000 square feet can be collected.

Polyculture:

Typically polyculture means farms growing two or more primary crops. Monocropping means growing just one crop. While a handful of primary crops can be sufficient, much greater biodiversity can be achieved through good design and gives results that prove to be more naturally sustainable as well as more productive. The addition of diverse plants that form positive symbiotic relationships, in what permaculture calls "guilds" (meaning communities of plants that work together ecologically), is the goal of good design and sustainability.

Pollination and Pollinators:

The world's pollinators are in crisis. Their population decline is taking place at a frightening rate. Most of us know that honey bees are often used to pollinate large monocrops, but wild bees and various other insects are indispensable pollinators as well.

As much as 80% of the world's food crops require pollination by insects. By putting pollinators at serious risk of extinction, so too do we put ourselves at serious risk of unprecedented hunger—even here in the First World. Habitat loss and disease contribute, but the main culprits are insecticides. There is very little to justify the use of insecticides. Large monocrops are about the only exception, but monocropping itself is so harmful to the environment that it is hard to defend. One reason monocropping is vulnerable to numerous pest problems is because even the smallest bit of habitat that might support pest controlling insects (that are often pollinators themselves) is removed.

Beneficial Insects:
The work of beneficial insects is to control pests, either as predators or as parasites. When farmers allow beneficial insects to replace pesticides, either by maintaining their natural habitat or by planting habitat for them, billions of dollars worldwide can be saved in crop protection. Most, if not all, beneficial insects are highly vulnerable to pesticides—often more so than the pests they were made to control.

There have been numerous examples worldwide showing that chemical insecticides actually increase pest populations, largely because they kill the beneficial insects that help control the insect pests.

It's been said that many insect pests find their desired food source by smell and that nearby aromatic plants can interfere with this ability by masking them with more potent scents. It has also been observed that pest insects avoid plants that harbor beneficial insects. Plants with insecticidal properties, which smell or taste bad to pests, might be useful for developing non-toxic pest repellants to protect vulnerable crops.

Weeds:
There are no strict rules regarding what is a weed and what is not. In ecology a weed is any plant that is not indigenous to the ecosystem in which it is found. In agriculture, a weed is any plant that is perceived as a non-crop and interferes with the agricultural system being practiced. For example, in industrial farming all plants other than the crop itself are weeds. Ironically, some plants that are eradicated as weeds are more nutritious than the crop being cultivated. They may even be improving the fertility of the soil.

Aside from hoeing and other hand-weeding methods used instead of toxic herbicides, there are other responsible ways to control weeds without poisons. Smother cropping is a well-established method. Using fast, aggressive-growing cover crops that outperform weeds can literally smother and eliminate them. Some cover crops like winter rye are allelopathic (in this case toxic) to weed seeds and can reduce weed populations by 50% during their first sowing without the support of other weed control strategies.

Mulching with cardboard covered with straw can control most weeds except those with starchy roots.

Living mulch is composed of weed suppressing plants that are grown with the plants to be protected. Annuals can also be used for a seasonal control, or perennials for a permanent planting.

Some of the worst weeds like bindweed, Canada thistle, and leafy spurge have been controlled with a vinegar spray, although it may need to be repeated for maximum control.

When large areas need to be weeded, sheep and goats will do the job. If the area is just a few acres or less, weeder geese, micro sheep, or micro goats are good choices. In all cases management is called for in order to avoid their munching your crops.

Weed control flamers using propane or radiant heat are very practical tools for the job. They both work by boiling the water in the plant cells until they burst. Repeated applications may be needed in some cases. Neither tool sets the plant on fire.

Herbicides made with plant oils are available, and some are organically certifiable. A few of these can control a variety of weeds including those with starchy roots.

What is called "biological control" typically uses insects to limit a single weed species. Insects that target just a single weed species and voraciously consume it are particularly effective if consumption includes the plant's reproductive system. Some insects, like the puncture vine weevil and the star thistle weevil, devour only the weed they are named for.

Wind Breaks and Shelter Belts:
In areas with severe problem winds, a properly designed windbreak offers significant benefits to agriculture. These benefits include a reduction in soil erosion, protection of crops from various mechanical injuries, increased crop yields, increased water efficiency, protection of livestock, improved outdoor conditions for humans, added habitat for wildlife, reduction of noise, protection for buildings, and increased indoor warmth in the winter.

Shelter belts also offer livestock shade on hot summer days and can provide forage for a longer period of time due to the reduction of moisture evaporation from the soil through shading.

Labor Intensive:
Every aspect of industrial agriculture is dependent on petrochemicals, a habit that must end if we are to survive on our planet. Much of this dependence fuels mechanization designed to minimize the need for labor. Labor intensive jobs on corporate industrial farms are in no way attractive jobs, but permaculture design and agroecology offer a way to make labor intensive farming jobs rewarding and gratifying. The demand for food from small farms will keep increasing in the coming decades. According to the United Nations Food Agency, agricultural food production must increase by 60% over the next few decades. This comes at a time when the majority of today's farmers are getting too old and will be

retiring soon. New farmers will need education on how to farm productively and ecologically.

Robotics is the currently favored labor-reducing technology, and it is advancing quickly. Then there are "Free Trade Agreements", like the North American Free Trade Agreement (NAFTA), which has cost the United States at least 500,000 jobs. The Korea-U.S. Free Trade Agreement (KORUS) has been reported to have cost the United States 100,000 jobs. When China joined the World Trade Organization (WTO), job loss in that country increased to over 3 million.

These and other trade agreements like the General Agreement on Tariffs and Trade (GATT) and the Multilateral Agreement on Investment (MAI) were never just about outsourcing jobs to benefit the 1%. The same is true of the Trans-Pacific Partnership (TPP), proposed for 2017. What they all have in common is the goal to create a global corporate plutocracy without environmental regulations or human rights provisions. In other words, to put an end to democracy and freedom.

What does this have to do with botanical treasures and the labor intensive model? First of all, today's economy operates by shipping food, raw materials, and products all over the world. In the future shipping will become more expensive and must decline. To counter this trend, we need to localize resources. As job loss in this country becomes epidemic and the cost of living skyrockets, grassroots movements may have to create their own economy to endure. Labor intensive means more jobs. High-yield, low-input agriculture that integrates permaculture, and agroecology in a worker-owned democratic co-op model could be one answer.

In this age of disenfranchisement, it could be highly empowering to form worker-owned democratic cooperatives to produce a variety of products. Not only would the members become a participant in a real democracy where their voice meant something, they would share the responsibilities as well as the rewards. Such co-ops promote a culture of sharing, equality, and unity.

With sufficient available land, a worker-owned democratic cooperative can farm a variety of raw materials and develop a host of cottage industries that produce a wealth of products. The complexity of our beautiful biosphere is based on biodiversity, and the co-op ecological paradigm is based on plant diversity, product diversity, and people diversity. Wealth becomes positive, protracted relationships rather than protracted material accumulation, meaning the world becomes a better place. *Botanical Treasures* is meant to contribute to such evolutionary modeling.

Farm Equipment:
Wherever possible, a no- or low-till practice is generally most desirable if it can be done organically. Even so, it can be hard to operate a farm without a tractor. Exceptions might be farms of ten acres or less with adequate investment in labor, or orchards or shelter belts that have been sown with a permanent perennial understory. Even then a tractor might be practical in certain cases.

Perhaps you remember the old international/harvester tractors with cat-tracks. Their weight was distributed evenly across the tracks to reduce the rutting and compaction that occurs with wheeled tractors. Compaction has an adverse effect on air and water penetration into the soil as well as the soil biota. Today, although uncommon, cat-track tractors are still being made.

Put away the destructive moldboard plow and replace it with the greatly superior chisel plow. The chisel plow maintains the all-important soil biota. It leaves organic matter on the surface rather than in the anaerobic subsoil where nitrogen is squandered. Left on the surface it decomposes much faster and without toxicity. Soil desiccation is minimized, capillary structure is maintained, and the soil surface is more resilient to wind and water erosion. At the same time, it plows deeper and faster and makes wider passes than a moldboard plow without causing plowpan.

Another tool that is both more functional and eliminates the need for additional implements (including the moldboard plow) is the disc harrow*. It does not plow at all, but simply chops up organic matter on the surface, thus accelerating decomposition by leaving it above ground just as nature does. At the same time, it breaks open the surface crust without adversely affecting the soil. Like the chisel plow it leaves the soil surface as it was, so that additional equipment to form a seed bed for planting is unnecessary. Use one or the other of these two implements, or use them both since they are actually complementary.

Oregon organic farmer Harry MacCormick recommends the Imants spader for small farms. You can get them for BCS walking tractors or for small to midsize tractors. They have a very slow rotary spade with a power harrow.

Glossary

General Terms:

alkali soil- soils excessively high in alkalinity that inhibits the growth of most plants due to a high percentage of exchangeable sodium.

alkalinity- any soil horizon with a ph value above 7.3.

allelopathic- the suppression of growth of one organism, plant or animal, by another due to the release of toxic substances.

anaerobic- a process that requires the absence of oxygen.

arroyo- a deep gully cut by an intermittent stream.

bosque- woodlands.

calcareous soil- a soil containing calcium carbonate.

coppice- trees and shrubs that re-sprout from the roots when cut down.

coppicing- a management practice that cuts back a woody plant at intervals of two or more years for repeated harvest.

culms- the stems of grasses and bamboos.

cultivar- indicates a cultivated variety or hybrid selected for cultivation.

cyme- a rather flat-topped flower cluster where the central flower is the first to open.

deciduous- trees and shrubs that shed all their leaves every year, the opposite of evergreens.

ecotypes- genetically distinct populations of individuals within a species that have adapted to climate, soils, topography, and other environmental factors.

entomologist- one who studies insects.

flocculate- the aggregation or clumping together of tiny individual soil particles, particularly fine clay, into granules, intentionally induced to improve soil aeration and drainage.

genus- a grouping of species in the same family believed to have evolved from a common ancestor.

humus- soil rich in carbon from decomposed (or nearly decomposed) organic matter, that is dark brown or almost black in color; humus rich soils are very fertile and contain important minerals.

hydrophobic- the tendency to repel or not mix with water.

hydrocarbons- a class of chemical compounds that contain only hydrogen and carbon, typically derived industrially from petroleum for developing various products like gasoline, plastics, and synthetic rubber, but also present in plants.

hydrophillic- the tendency to mix with water.

hyperaccumulators- plants that are rich in minerals that are low in the topsoil in which they grow; it is thought that the roots of such plants reach down into the subsoil until they find minerals lacking in the topsoil; in death or defoliation these minerals enrich the topsoil.

inflorescence- the flowering part of a plant; for example a group of individual flowers on a single stem.

intercropping- a reference to two or more crops grown together.

key- dry one-seeded fruit with a wing (samara).

keystone- species, groups of species, habitats, or abiotic factors such as wildfire that are critical to an ecosystem's processes.

limestone- limestone soils are calcareous and are often found in arid climates; limestone is used to neutralize acid soils.

mineral accumulators- any of a number of plants that absorb and contain high levels of various minerals; whenever they defoliate or die, these minerals are released in the soil.

nitrogen fixing plants- plant species that have a symbiotic relationship with nitrogen fixing bacteria that attach to their roots.

node- the location of a stem where one or more leaves are attached, sometimes marked by a distinct scar, a swelling, or a ring.

ovary- the part of a flower that bears the ovule that when fertilized develops a seed.

perfect flower- a flower having both functional stamens and pistils; self-fertile.

pH- a numerical measurement of the acidity or alkalinity of soils with a scale form 0 to 14; a pH of 7.0 indicates neutral, lower numbers indicate the degree of acidity, higher numbers indicate the degree of alkalinity; the best soils are between 6.5 and 7.0.

pollard- prune top and upper branches to encourage new growth.

pyrolysis- the decomposition of a substance by heating it to a high temperature in the absence of air.

raceme- a branchless, elongated group of flowers; each flower borne has its own stock.

rhizobia- nitrogen fixing bacteria that attach to the roots of many legumes.

rhizome- a modified stem that grows horizontally under the soil.

riparian- refers to the banks and flood plains of rivers and streams.

ruminant- animals such as cattle, goats, and sheep with a four-chambered stomach; food is stored and fermented then re-eaten.

saline- a soil with enough soluble salts to inhibit plant growth and productivity, typically evidenced by a white salty crust on the soil surface.

scarification- the process of soaking hard-coated seeds in hot water before sowing to encourage germination.

species- a specific type in a genus.

stigma- the part of a female that receives male pollen.

stolon (Stoloniferous)- a horizontal stem that spreads above or below ground and roots at the tip to produce a new plant.

sucker- a stem produced by a spreading root or rhizomes; they commonly occur in groups of a few to many, may sprout far out from the mother plant, and may be forced to sucker by heavy pruning.

sucrose- a disaccharide sugar with two molecules, one glucose and the other fructose (monosaccharide).

surfactant- a substance that when added to a liquid, like water, increases wetness over a wider area.

taxonomist- one who classifies living organisms.

Texas root rot- a fungal disease located in the arid southwestern United States at altitudes below 3,500 feet.

thermoplasticity- becoming soft with heat and becoming hard with cooling.

trans polyisoprene- organic molecule found in natural sources of rubber.

umbel- an individual flower stalk with a group of flowers radiating from a single axis; commonly umbrella-like or flat-topped.

variety- a minor variation of a basic botanical species as it occurs in nature, like a subspecies; the term is also applied to differences resulting from cultivation as in cultivars.

winnow- a method to separate the chaff from the seed using air currents.

Medical Terms:

adaptogen- an herb that safely increases resistance to stress; adds balance and resilience to the body's functions.

alkaloids- a large group of compounds containing nitrogen and generally found in plants; although potentially toxic, some are used in medicine; morphine is one example.

alternative- a medicinal herb or substance that progressively restores health and vitality to specific bodily functions.

analgesic- a pain relieving substance.

antiarthritic- an agent that can provide relief from arthritis and gout.

antibacterial- a substance that inhibits the growth or spread of bacteria.

antibiotic- an agent that inhibits or kills harmful microbes.

anticatarrhal- helps remove excess mucus from the body.

antifungal- a substance that controls or inhibits the growth or spread of fungi.

antihemorrhagic- an agent that prevents or controls hemorrhage or bleeding.

antihepatoxic- an agent that protects liver cells from chemical damage.

antihypertensive- a substance that lowers blood pressure.

antihyperlipidemic- helps reduce lipid concentrations in the blood stream.

anti-inflammatory- a direct reduction or neutralizing of inflamed tissue.

antimutagenic- an agent that prevents or combats genetic mutations.

antioxidant- an agent that scavenges free radicals and prevents oxidation.

antipyretic- reduces fever.

antirheumatic- an agent that prevents or relieves rheumatism.

antisclerotic- helps prevent the hardening of tissue.

antiseptic- compounds that check or prevent decay and purification caused by microbes.

antispasmodic- prevents or relieves spasms of the smooth muscle tissue.

antithrombotic- prevents blood coagulation (thrombosis).

antiviral- checks growth of viral pathogens.

apoptosis- programmed cell death; a genetic limit on a cell's lifespan.

astringent- an agent that constricts or binds.

carminative- promotes good digestion, prevents or relieves gas, and reduces inflammation of the gastrointestinal tract.

cathartic- a purgative that produces bowel movements; stronger than laxatives.

coumarins- compounds that inhibit blood clotting.

cyclooxygenase inhibitors- agents that control or inhibit enzymes such as Cox 1 and Cox 2 that play a central role in inflammatory diseases and other conditions.

demulcent- a mucilaginous herb that soothes of softens irritated or inflamed tissues, particularly the mucus membranes.

diaphoretic- an agent that promotes perspiration to eliminate body wastes.

diuretic- increases urination.

emetic- a substance that promotes vomiting..

emollient- a substance that soothes and softens external tissue.

enzymes- proteins produced by living cells that act as catalysts for chemical reactions.

essential fatty acids- polyunsaturated fats that are indispensable for normal growth and healthy arteries and nerves; the body cannot produce them so they must come from the diet.

expectorant- an agent that promotes the expulsion of mucus from the respiratory tract, soothing dry irritating coughs and bronchial spasm.

febrifuge- a substance that reduces fever.

flavonoid- plant pigments that produce a significant diversity of physical influences in the body.

furanofuran- a lignin found in Eucommia ulmoides and the Fraxinus species that acts as an antihypertensive by inhibiting cAMP activity.

glycemic index- a guide to how much various foods raise blood glucose levels.

glycoside- through hydrolysis yields a sugar (glucose) and another molecule; many kinds of glycosides are common in a great number of plants, numerous glycosides contain medicinal properties.

hemostatic- an agent that checks bleeding.

hepatic- pertaining to the liver.

hypocholesteremia- decreased blood cholesterol at undesirable levels.

hypoglycemic- an agent that lowers abnormally elevated blood sugar.

hypotensive- an agent that lowers abnormally elevated blood pressure.

immunostimulant- an agent that invigorates the immune system.

mucopurulent- consisting of mucus and pus.

parasiticidal- an agent that kills parasites.

phytochemical- any one of hundreds of plant-based chemicals, many of which provide important functions and benefits for the human body.

poultice- the application of a soft, moist substance such as medicinal herbs, to a surface injury to promote healing and relieve pain.

prebiotics- an indigestible ingredient in certain foods that promotes the growth of beneficial colonic microflora.

probiotics- beneficial bacteria that favorably affect the balance of the intestinal microflora, aiding good digestion, providing resistance to infection, inhibiting harmful microbes, and boosting the immune system.

refrigerant- reduces fever by promoting perspiration.

resinoid- a soft, aromatic resinous plant material.

saponin- a non-absorbable glucoside found primarily in the roots of some plants, it can produce a lather, and act as a soap.

sesquiterpine lactones- inhibit platelet aggregation and the release of serotonin from platelets; highly irritating to the eyes, nose, and gastrointestinal tract.

sudorific- an agent that produces perspiration.

tincture- a medicinal herb preparation commonly found in an alcohol base, but also in vegetable glycerin.

tonic- a substance that strengthens and restores; used as a disease preventative, particularly for chronic conditions.

vasodilator- an agent that dilates blood vessels.

vulnerary- an agent applied to heal internal and external wounds.

Bibliography

Adrosko, Rita J., *Natural Dyes and Home Dyeing*, Dover Publications, 1971.

Agroforestry Research Trust, *Agroforestry News (Medicinal Shrub Crops)*, Vol. 8. No. 3. April, 2000.

Agroforestry Research Trust, *Agroforestry News (Fiber Sources from Bark)*, Vol. 13 No. 3. May, 2005.

Agroforestry Research Trust, *Agroforestry News (The Myrtles)*, Vol. 14 No. 1, November, 2005.

Airola, Paavo (Ph.D./N.D.), *The Miracle of Garlic*, Health Plus Publishers, 1978.

Ajilvsgi, Geyata, *Wild Flowers of Texas*, Shearer Publishing, 1984.

Anderson, M. Kat, *Tending the Wild (Native American Knowledge and the Management of California's Natural Resources)*, University of California Press, 2005.

Bailey, Ethel Zoe and Liberty Hyde Bailey, *Hortus Third (A Concise Dictionary of Plants Cultivated in the United States and Canada)*, revised and expanded by the staff of Liberty Hyde Hortorium, MacMillan Publishing Co. Cornell University, 1976.

Better Nutrition, *Oregano*, April, 2002.

Blanchan, Neltje, *The Nature Library – Wild Flowers*, 1917, Doubleday and Co., Inc. 1926.

Blume, David, *Alcohol Can Be a Gas (Fueling an Ethanol Revolution for the 21st Century)*, International Institute for Ecological Agriculture, 2007.

Bremness, Lesley, *Eyewitness Handbook—Herbs*, Dorling-Kidersley, London, 1994.

Brenzel, Kathleen N. (editor), *Sunset Western Garden Book*, Sunset Publishing Corp., 2001.

Carr, Anna, *Good Neighbors (Companion Planting for Gardeners)*, Rodale Press Inc. 1985.

Chamberlin, Susan, *Hedges, Screens, and Espaliers*, HP Books, 1983.

Chonxi, Yue and Steven Foster, *Herbal Emissaries – Bringing Chinese Herbs to the West*, Healing Arts Press, 1992.

Coate, Barrie (Principal Author and Technical Consultant), *Water Conserving Plants and Landscapes for the Bay Area*, second edition, East Bay Municipal Utility District, 1990.

Couplan, Francois (Ph.D.), *The Encyclopedia of Edible Plants of North America*, Keats Publishing, 1998.

Curtin, L.S.M, *Healing Herbs of the Upper Rio Grande*, Southwest Museum of Los Angeles, California. 1965.

Dannen, Kent and Donna Kent, *Rocky Mountain Wildflowers*, Tundra Publications, 1981.

Dastur, J.F. (FNI), *Useful Plants of India and Pakistan (Authoritative Work on Trees and Plants of Industrial, Economic, and Commercial Utility)*, 1964, D.B. Taraporevala Sons and Co. Private LTD, 1985.

Desert Botanical Garden Staff, *Arizona Highways Presents Desert Wildflowers*, Arizona Department of Transportation, 1988.

Deur, Douglas and Nancy J. Turner, *Keeping It Living (Traditions of Plant Use and Cultivation on the Northwest Coast of North America)*, University of Washington Press, 2005.

Dirr, Michael A., *All About Evergreens*, Ortho Books, 1984.

Dirr, Michael A., *Manual of Woody Landscape Plants (Their Identification, Ornamental Characteristics, Culture, Propagation, and Uses)*, Stipes Publishing Company, 1983.

Dodge, Natt N., *Flowers of the Southwest Deserts*, Southwest Parks and Monuments Association, 1985.

Duffield, Mary and Warren D. Jones, *Plants for Dry Climates (How to Select, Grow, Enjoy)*, H.P. Books, 2001.

Duke, James A. (Ph.D.), *Handbook on Edible Weeds*, CRC Press, 1992.

Duke, James A. (Ph.D.), *The Green Pharmacy*, Rodale Press, 1997.

Duke, James A. (Ph.D.), *The Green Pharmacy Guide to Healing Foods*, Rodale Press, 2008.

Duke, James A. (Ph.D.) and Steven Foster, *Medicinal Plants and Herbs Eastern/Central*, "Peterson Field Guides," Houghton Mifflin Co. 2000.

Dunmire, William W. and Gail Tierney, *Wild Plants of the Pueblo Province (Exploring Ancient Enduring Uses)*, Museum of New Mexico Press, 1995.

Dunmire, William W. and Gail D. Tierney, *Wild Plants and Native Peoples of the Four Corners*, Museum of New Mexico Press, 1997.

Ebeling, Walter, *Handbook of Indian Foods and Fibers of Arid America*, University of California Press, 1986.

Elias, Thomas S. and Peter. A. Dykeman, *A Field Guide to North American Edible Wild Plants*, Outdoor Life Books, 1982.

Everett, Thomas H., *Living Trees of the World*, Doubleday and Company, Inc., 1968.

Facciola, Stephan, *Cornucopia II (A Source Book of Edible Plants)*, Kampong Publications, 1998.

Fern, Ken, *Plants for a Future*, Permanent Publications, England, 2000.

Foster, Steven and Christopher Hobbs, *Western Medicinal Plants and Herbs*, Peterson Field Guides, The Houghton Mifflin Company, 2002.

Gardner, Joann, *Living With Herbs*, The Countryman Press, 1997.

Genders, Roy, *Edible Wild Plants (A Guide to Natural Foods)*, a co-publication of EMB-service for Publishers and Van Dermark Editions (U.S. edition), 1988.

Gibbons, Euell, *Stalking the Wild Asparagus*, David McKay Co., Inc. 1962.

Grieve, M., *A Modern Herbal (The Medicinal, Culinary, Cosmetic, Economic Properties, Cultivation, and Folklore of Herbs, Grasses, Fungi, Shrubs, and Trees With Their Modern Scientific Uses)*, Vol. 1. Dover Publications, Inc. 1971.

Haigh, Charlotte, *The Top 100 Immunity Boosters*, Duncan Baird Publishers, 2005.

Harrington, H. D., *Edible Native Plants of the Rocky Mountains*, University of New Mexico Press, 1967.

Hart, Jeff, *Montana-Native Plants and Early Peoples*, The Montana Historical Society, 1976.

Haskin, Leslie L., *Wild Flowers of the Pacific Coast*. 1967, Binfords and Mort Publishers, 1970.

Herbs For Health, *Herbs to Watch – Oregano*, March-April, 1999.

Hibler, Janie, *The Berry Bible*, William Morrow/Harper Collins Publishers, 2004.

Hill, Ray, *Propolis a Natural Antibiotic*, Thorson Publishers Limited. June, 1979.

Hoerr, Normand L. (M.D.) and Arthur Osol (Ph.D.), editors, *Blakiston's Illustrated Pocket Medical Dictionary*, McGraw-Hill, 1952.

Hoffmann, David (FNIMH/AHG) *Medical Herbalism (The Science and Practice of Herbal Medicine)*, Healing Arts Press, 2003.

Hottes, Alfred C., *The Book of Shrubs*, A.T. De la Mare Co., Inc. 1952.

Hottes, Alfred C., *The Book of Trees*, fourth printing, A.T. De la Mare Co., Inc. 1947.

Hunter, Beatrice T., *Gardening Without Poisons*, 1964, second edition, Berkeley Publishing Corp. 1971.

In Good Tilth, *Research Reports "Unusual Feed Supplement Could Ease Greenhouse Gassy Cows,"* Vol. 21 #5, November-December 2010.

Jaeger, Edmund C., *Desert Wild Flowers*, 1940, Stanford University Press, 1969.

Kearney, Thomas H., Robert H. Peebles, et al. *Arizona Flora*. University of California Press, Berkeley and Los Angeles. 1951.

Kindscher, Kelly, *Medicinal Wild Plants of the Prairie (An Ethnobotanical Guide)*, University Press of Kansas, 1992.

Kirk, Donald R., *Wild Edible Plants of Western North America*, Naturegraph Publishers, Inc. 1970, 1975.

Kirschman, John D., *Nutrition Almanac*, sixth edition, McGraw-Hill Books, 2007.

Lewington, Anna, *Plants For People*, Oxford University Press, Inc. 1990.

Lininger Jr., Schuyler (W.-D.C.), Alan Gaby (R-M.D.), Steve Austin (N.D.), Donald Brown (J-N.D.), Jonathon Wright (V-N.D.), and Alice Duncan (DC/CCH), *The Natural Pharmacy*. "Prima Health," Prima Communications, Inc. and Health Notes, Inc. 1999.

Lust, Terresa, *Summer Gems*, article in the Herb Companion August-September, 2000.

Mabberley, D.J., *The Plant Book (A Portable Dictionary of the Vascular Plants)*, second edition, Cambridge University Press, 1997.

Maybey, Richard, *Plantcraft (A Guide to the Everyday Use of Wild Plants)*, Universe Books, 1977.

Medsger, Oliver Perry, *Edible Wild Plants (The Complete Authoritative Guide to Identification and Preparation of North American Edible Wild Plants)*, 1939, MacMillan Publishing Co., Inc. 1966.

Miller, Richard A., *Native Plants of Commercial Importance*, Oak, Inc. 1988.

Moore, Michael, *Medicinal Plants of the Mountain West*, Museum of New Mexico Press, 1979.

Moore, Michael, *Medicinal Plants of the Desert and Canyon West*, Museum of New Mexico Press, 1989.

Mowrey, Daniel B. (Ph.D.), *The Scientific Validation of Herbal Medicine*, Keats Publishing, Inc. 1986.

Murray, Michael (N.D.) and Joseph Pizzorno (N.D.), *Encyclopedia of Natural Medicine*, second edition, Prima Publishing, 1998.

Murray, Michael (N.D.), Joseph Pizzorno (N.D.), and Lara Pizzorino (M.A./I.M.T.), *The Encyclopedia of Healing Foods*, Atrai Books, 2005.

Munz, Philip A., *California Desert Wildflowers*, 1962, University of California Press, 1975.

Munz, Philip A. and David D. Keck, *A California Flora and Supplement*. 1959. University of California Press. 1968.

Nabhan, Gary Paul, *Gathering the Desert*, third printing, University of Arizona Press, 1987.

Niethammer, Carolyn, *American Indian Foods and Lore*, Collier MacMillan Publishers, 1974.

Noad, Tim C. and Ann Birnie, *Trees of Kenya*, published by T.C. Noad and A. Birnie, 1989.

Palmer, E. Laurence (Ph.D), *Fieldbook of Natural History*, second edition, E.P. Dutton and Company, Inc. 1957.

Pederson, Mark, *Nutritional Herbology*, 1987, Wendell W. Whitman Co. revised and expanded 1998.

Pellet, Frank C., *American Honey Plants*, fifth edition, Dadant and Sons, Inc. 1976.

Perry, Frances, editor, *Simon and Schuster's Complete Guide to Plants and Flowers*, Simon and Schuster, 1974.

Peterson, Lee Allen, *Edible Wild Plants of Eastern and Central North America*, Roger Tory Peterson Field Guides, Eaton Press, 1977.

Phillips, R. and N. Foy, *Random House Book of Herbs*, Random House Publishers, 1990.

Pizzetti, Mariella and Paola Lanzara, English translation by Hugh Young, *Guide to Trees (A Field Guide to Conifers, Palms, Broadleafs, Fruits, Flowering Trees, Trees of Economic Importance)*, Simon and Schuster, 1977.

Pojar, Jim and Andy MacKinnon, editors, *Plants of the Pacific Northwest Coast (Washington, Oregon, British Columbia, and Alaska)*, Lone Pine Publishing, 1994.

Polunin, Oleg, *A Field Guide – Flowers of Europe*, Oxford University Press, 1969.

Quist, John A., *Urban Insect Pest Management (For Deciduous Trees, Shrubs, and Fruit)*, second printing, Pioneer Science Publications, 1981.

Randall, Warren R., Robert F. Keniston, and Dale N. Bever, *Manual of Oregon Trees and Shrubs*, reprint by Oregon State University, 1976.

Rateaver, Bargyla and Gylver Rateaver, The Rateavers, United States, 1993.

Reich, Lee, *Landscaping With Fruit*, Storey Publishing, 1992.

Reid, Daniel P., *A Handbook of Chinese Healing Herbs*, Shambala Publications, Inc. 1995.

Reid, Daniel P., *Chinese Herbal Medicine,* Shambala Publications, Inc. 1987.

Report of an Ad Hoc Panel of the Advisory Committee on Technology Innovation Board on Science and Technology for International Development Commission on International Relations National Research Council, *Tropical Legumes (Resources for the Future),* National Academy of Sciences, 1979.

Riotte, Louise, *Secrets of Companion Planting for Successful Gardening,* Garden Way Publishing, 1975.

Robson, Kathleen A., Alice Richter, and Marianne Filbert, *Encyclopedia of Northwest Native Plants for Gardens and Landscapes,* Timber Press, 2008.

Saunders, Charles Francis, *Edible and Useful Wild Plants of the United States and Canada,* 1920, Dover Publications, Inc. 1976.

Shih-Chen, Li, English translation by F. Porter Smith (M.D.) and G. A. Stuart. (M.D.), *Chinese Medicinal Herbs,* Pen-Tsao (Materia Medica), 1578. Georgetown Press, 1973.

Simmons, Alan F., *Growing Unusual Fruit,* Walker and Company, 1972.

Smith, Joshua, *A Few Good Plants (The Easy to Grow Food Pharmacy),* unpublished, 2000.

Smith, J. Russell, *Tree Crops (A Permanent Agriculture),* The Devin-Adair Company, 1950.

Sturdivant, Lee and Tim Blakely, *Medicinal Herbs in the Garden, Field, and Marketplace,* San Juan Naturals, 1999.

Sturtevant, E. Lewis (Ph.D.). Author. U.P. Hendrick, editor, *Sturtevant's Edible Plants of the World,* originally titled *Sturtevant's Notes on Edible Plants* published in 1919, Dover Publications unabridged and unaltered republication, 1972.

Sweet, Muriel, *How Trees Help Your Health,* Naturegraph Publishers, 1965.

Teeguarden, Ron, *Chinese Tonic Herbs,* Japan Publications, Inc. Seventh Printing, 1992

Tillotson, Allen K. (Ph.D./A.H.G./D.AY.), Nai-Shing Hu Tilloston (O.M.D., L.Ac.) and Robert. Abel Jr. (M.D.), *The One Earth Herbal Sourcebook,* Kensington Publishing Corp. 2001.

Turner, Mark and Phyllis Gustafson, *Wildflowers of the Pacific Northwest,* Timber Press Inc, 2006.

U.S. Department of Agriculture, *Yearbook of Agriculture 1992 (New Crops, New Uses, New Markets, Industrial and Commercial Products from U.S. Agriculture),* 1992.

U.S. Department of Agriculture, *Yearbook of Agriculture 1950-1951 (Crops in War and Peace).*

U.S. Department of Agriculture (prepared by the Forest Service), *Range Plant Handbook,* 1937.

Venes, Donald (M.D./M.S.J.), editor, *Taber's Cyclopedic Medical Dictionary,* F.A. Davis Co. 2001.

Vines, Robert A., *Trees, Shrubs, and Woody Vines of the Southwest,* 1960, University of Texas Press, Austin, 1986.

Vogel, Virgil J., *American Indian Medicine,* first edition, University of Oklahoma Press, 1970.

Wasowski, Sally and Andy Wasowski, *Native Texas Plants,* Texas Monthly Press, 1988.

Wetherbee, Kris, *Oregano the Perfect Tomato Partner,* Organic Gardening, August-September, 2005.

Willard, Terry (Ph.D.), *Edible and Medicinal Plants of the Rocky Mountains and Neighbouring Territories,* Wild Rose College of Natural Healing, LTD. 1992.

Wren, R. C. (F.L.S.), *Potter's Encyclopedia of Botanical Drugs and Preparations,* sixth edition, Potter and Clark, LTD., 1950.

Yepsen Jr., Roger B., *Trees for the Yard, Orchard, and Woodlot (Propagation, Pruning, Landscaping, Orcharding, Sugaring, Woodlot Management, Traditional Uses),* Rodale Press, Inc. 1976.

Author's note: The above list of references contains all those that contributed the most to this book, as well as some of minor significance to this work. Others of the latter category have been omitted.

Illustration Sources

Cover photo- Asclepias speciosa, permission granted by Thomas Checkley, http://checkconnect.net/blog/plants-of-our-hikes/flowers-of-our-hikes/.

Page 2- Achillea millefolium L., W. H. Fitch, *Illustrations of the British Flora* 1924, from Creative Commons.

Page 16- Aleurites moluccana, C. Martius, A.G. Eichler, I. Urban, *Flora Brasiliensis*, vol. 11(2): fasicle 64, t. 45, 1874, permission granted by Peter H. Raven Library/Missouri Botanical Garden, St. Louis, MO.

Page 23- Allium sativum, Q.P. van der Meer and Anggoro H. Permadi, 1993, J.S. and Kasem Piluek (Editors). *Plant Resources of South-East Asia* No 8. Vegetables. Pudoc Scientific Publishers, Wageningen, Netherlands. pp. 77–80, from Creative Commons.

Page 41-Asclepias speciosa, pen and ink by Lesley Randall, permission granted by Lesley Randall.

Page 62- Cucurbita foetidissima, from Arizona-Sonora Desert Museum website, https://www.desertmuseum.org/.

Page 69- Eucommia ulmoides, Feng Zhonguan, redrawn by Cai Shuqin, from www.efora.cn.

Page 78- Fraxinus americana, C.S. Sargent, *The Silva of North America*, vol. 6: t. 268 (1892) [C.E. Faxon], permission granted by Peter H. Raven Library/Missouri Botanical Garden, St. Louis, MO.

Page 90- Gledistsia triacanthos, William C. Grimm, The Book of Trees, The Telegraph Press, Harrisburg, Pennsylvania, 1831, from Creative Commons.

Page 101-Lespedeza Bicolor L. H. Bailey *Standard Cyclopedia of Horticulture*, New York, New York: The MacMillan Company, 1917, permission granted by Florida Center of Instructional Technology College of Education Univ. of South Florida, image from etc.usf.edu/clipart.

Page 105- Myrtis Communis, Prof. Dr. Otto Wilhelm Thomé, *Flora von Deutschland, Österreich und der Schweiz*, Gera, Germany, 1885, from Wikimedia Commons.

Page 111- Origanum Vulgare L., W. Woodville, *Medical Botany*, vol. 3: t. 164, 1793, permission granted by Peter H. Raven Library/Missouri Botanical Garden, St. Louis, MO.

Page 123- Phragmites Communis, *Atlas des plantes de France*, A. Masclef, 1891, from Wikimedia Commons.

Page 133- Prosopis glandulosa- from Stanford University website, https://trees.stanford.edu/ENCYC/PROgla.htm

Page 151- Rosa Species, *Botanical Register*, vol. 5: t. 425 (1819) J. Lindley, permission granted by Peter H. Raven Library/Missouri Botanical Garden, St. Louis, MO.

Page 177- Sambucus Species- C.S. Sargent, *The Silva of North America*, vol. 5: t. 221, 1892, [C.E. Faxon], permission granted by Peter H. Raven Library/Missouri Botanical Garden, St. Louis, MO.

Page 202- Trachycarpus Fortunei- D. Maraval, from AMAP Studio website, http://amapstudio.cirad.fr/soft/principes/page_trachycarpus

Page 206- Typha latifolia, permission granted by University of Florida Center for Aquatic and Invasive Plants, CAIP-website@ufl.edu

Page 226- Yucca aloifolia, C.S. Sargent, *The Silva of North America*, vol. 10: t. 502 (1898) [C.E. Faxon], permission granted by Peter H. Raven Library/Missouri Botanical Garden, St. Louis, MO.

Page 245- Ziziphus jujuba, Francisco Manuel Blanco (O.S.A.), *Flora de Filipinas*, Gran edicion,[Atlas I, from Wikimedia Commons.

Page 258- Hardiness Map of United States and Canada, from *All About Evergreens*, Michael Dirr, Ortho Press, 1984.

The Author

Joshua Smith has been designing and installing permaculture/agroecological projects that mimic nature for forty years. He has also taught permaculture throughout the west, including having taught with Bill Mollison, the founder of permaculture. Joshua helped start Seeds of Change, one of the first all organic seed companies in the United States. He worked as a permaculture designer for the Sol y Sombra Foundation in Santa Fe, New Mexico and for the Malachite Farm School in Gardner, Colorado. He was the director of permaculture and ecoforestry at Earth Star Farm in Boulder, Colorado and was a permaculture design instructor at Naropa University, Boulder, Colorado. Joshua has also practiced ecoforestry for over thirty years.

Made in the USA
Lexington, KY
06 March 2017